Sky in a Bottle

Sky in a Bottle

Peter Pesic

The MIT Press
Cambridge, Massachusetts
London, England

First MIT Press paperback edition, 2007

MIT Press books may be purchased at special quantity discounts for business or sales promotional use. For information, please email special_sales@mitpress.mit.edu.

This book was set in Stone Serif and Stone sans by SNP Best-set Typesetter Ltd., Hong Kong, and was printed and bound in the United States of America.

Library of Congress Cataloging-in-Publication Data

Pesic, Peter.
Sky in a bottle / Peter Pesic.
p. cm.
Includes bibliographical references and index.
ISBN 978-0-262-16234-0 (hc : alk. paper)—978-0-262-66200-0 (pb)
1. Science—History. 2. Physics—History. 3. Science—Popular works. 4. Inquiry (Theory of knowledge). I. Title.

Q162.P37 2005
500—dc22

2005041578

Permission for use of the figures has been kindly given by the following: Springer-Verlag (figs. 3.6a, 5.1, 5.4); the President and Fellows of Harvard College: from HOLLIS 002102533 (fig. 4.1); Rayleigh Archives, Air Force Research Library (fig. 6.3); Jan Sengers and the American Chemical Society (fig. 8.3); Eugene Clothiaux (fig. 10.3); Syndics of Cambridge University Library (fig. 10.5).

Thanks to Eugene Clothiaux, Judy Feldmann, Chryseis Fox, John Grafton (Dover Publications), John Howard, Sara Meirowitz, Alexei Pesic, Andrei Pesic, and Ssu Weng, for their help with the figures.

10 9 8 7 6 5 4 3

For Andrei and Alexei

One of these mornings
You're going to rise up singing
Then you'll spread your wings
And you'll take to the sky

Contents

Introduction

Imagine a summer afternoon with a serene sky. Its color is subtly variable, at times azure or almost indigo. But why is the sky *blue*?

Though a child could pose this question, it also puzzled Plato, Leonardo, and even Newton, who unlocked so many other secrets. Centuries passed; the question remained unresolved. As late as 1862, Sir John Herschel listed the color and polarization of skylight as "the two great standing enigmas of meteorology."[1] Solving these mysteries required powers of deduction that would have taxed Sherlock Holmes. Like a compelling detective story, this tale has many curious turns and surprises. The clues seem simple, but they force us to reexamine what seems obvious. The characters include some remarkable human beings. The identity of the culprit is especially revealing. What is more, the story is not over; we still do not fully understand why we see the sky as we do.

Best of all, the mystery is in the very air around us. Trying to deal with its elusive subtlety leads to a further question: Can we put the sky in a bottle? That is: Can we recreate that blue on earth, not just contemplate it in the vast expanse of the atmosphere? If we could bottle the sky, we might understand its blueness, as if now the prime suspect were our captive, subject to probing interrogation. This new perspective invites us to consider other mysteries, especially why the night sky is as dark (or bright) as it is.

The quest to put the sky in a bottle led to important realizations. I will try to present some new insights, as well as give new expression to others that call for restatement. I have tried to make this story as accessible as possible, with references and additional technical or mathematical notes at the end of the book, to help those who wish to go further. My goal was to bring the deep underlying questions to light in a thoughtful and faithful way, without oversimplification. Thus, I hope that this book can be read both as an engrossing tale and as a serious history. For those who would like to try some simple experiments, the text will refer to relevant passages in appendix A. Appendix B contains several hitherto unpublished letters between some of the protagonists.

Each episode in this story contains its own special grain of truth, fascinating in its own right, not merely as a tale of anticipations and false starts but of thought working *in extremis*. The cast of characters is large, though some will take the stage only for a moment. Though I will offer some human detail for a few personages, I could not do so for all without hindering the larger narrative, whose protagonist is not a person but a question. Such a story has a special dynamic that is human but also something more.

The quest to understand the blue sky and put it in a bottle opens larger perspectives, both scientific and humanistic. The color and brightness of the sky touches the secrets of matter and light, the scope of the universe in space and time, the destiny of the earth, and deep human feelings.

1 Out of the Blue

A mystery must emerge before it can be solved. Thousands of years may have passed before someone wondered about the sky's color. Questions tend to accumulate more around unfamiliar, strange happenings than around the everyday and ordinary. Yet unasked questions also point to fundamental beliefs and prohibitions.

Though their civilizations were very different, both ancient Greece and China tended not to speak about the sky's color. In the ancient Chinese *Book of Songs* (*Shi Jing*), the phrase *cang tian* denotes vast or azure heaven, the overarching and immense vault of the sky and by implication the Power ruling it. The songs address the sky with yearning, calling to a realm far above human suffering, as in this lament over the old capital of the state of Zhou and the king who lost it:

Blue Heaven far, far above me,
What kind of man is he?[1]

Here, the impassive remoteness of the blue sky contrasts with human suffering and loneliness.

For the ancient Greeks and Chinese, azure had distinctly inhuman connotations, because death changes the healthy red color of the body to an unearthly blue, cyanosis. At Fengdu, in the main gorge of the Yangzi river, stands a temple reputed by Taoist tradition to be the gate of Hell, the entrance to the afterlife that must be traversed by dead souls.[2] The gates are painted a garish sky blue, not to be touched by the living, at peril of premature death. In contrast, yellow was the imperial color, reserved for the emperor's use, and red the color of life and celebration, used even today for marriages. Yet in India the divinities were often portrayed with dark blue skins and the poet Kālidāsa celebrated a sky "as dark blue as a sword."[3]

Purple was the imperial color of Rome, but blue was definitely the color of the barbarians. The ancient Britons dyed their bodies blue, a frightening sight to enemies who associated this color with death; the Roman historian Tacitus described the blue-daubed Britons' "spectral army." British women injected blue dye beneath the skin. In contrast, the Romans wore blue to denote mourning and considered blue eyes a kind of deformity, a sign of bad character, a barbarian trait.[4] Likewise, in Arabian folklore, blue eyes were sometimes considered signs of the evil eye. According to the twelfth-century Persian poet Farid ud-din Muhammad Attar, "heaven bears the blue color of sorrow as a sign of mourning that it has not attained the goal of its search to know the essence of God."[5]

For the ancients, the heavenly realm was divine. Because of its superlative power, the sacred could be dangerous, hence better left unuttered. If so, it would be better not to name the sky's color or touch its alien power. Recounting the Greek creation stories, Hesiod tells that Sky (*Ouranos*) was the first child of Earth, "equal in size with herself, to cover her on all sides."[6] The poet Aeschylus tells that "pure sky [*Ouranos*] desires to penetrate the earth, and the earth is filled with love so that she longs for blissful union with the sky. The rain falling from the beautiful sky impregnates the earth, so that she give birth to plants and grain for beasts and men."[7] Among the descendants of this primal union was the high god Zeus, dwelling in the sky with the other Olympian gods. The deities identified with the earth stood apart from those of the sky.

Even now, *heaven* connotes a divine realm radically different from earth, whereas *sky* is a more neutral term. Modern science turned *heaven* into *sky* and then *atmosphere*. Yet we continue to refer to "the heavens," showing how deeply rooted is the ancient conception. The Greeks contrasted the heavens, *ouranos*, with what they called *physis*, a realm of growth and change extending between the earth and the moon. Our word "physics" typically refers to inanimate matter, but the Greek word *phyein* means to grow and change as living organisms do. Though we moderns characteristically think of the earth as just another planet in space, the Greek words *ouranos* and *physis* indicated utterly different realms. Of the two, *ouranos* was nobler, the realm of celestial, immortal bodies. *Physis* is the mortal realm where we are born, grow, and die.

Among the earliest Greek thinkers, Parmenides gave visionary form to the search to discern something constant and unchanging behind the manifold flux of the world. In his poem, an unnamed goddess sets forth the quest on which we now embark:

> Gaze steadfastly at things which, though far away, are yet present to the mind. . . . On the one hand, there is the fire of the upper sky [*aether*], gentle, rarefied, and everywhere identical with itself; on the other hand, there lies opposed to it utter darkness, dense and heavy. . . . You shall come to know the nature of the sky [*aether*], and the signs of the sky, and the unseen works of the pure bright torch of the sun and how they

came into being. . . . You shall know also the encompassing heaven [*ouranos*], whence it arose, and how Necessity grasped and chained it so as to fix the limits of the stars.[8]

Since the heavens transcend terrestrial phenomena, special words describe their substance, as opposed to the air we see around us. *Aether* literally means ever-running or ever-blazing, characterizing the ceaseless radiance of the celestial realm. Where *ouranos* denotes deep heaven and *aēr* the lower air, close to the earth, aether is the upper realm of the atmosphere, the domain of clouds and of Zeus. Thus, one of the Golden Verses of the Pythagorean brotherhood tells that "when after divesting yourself of your mortal body you arrive in the pure upper aether, you will be a god, an immortal, incorruptible; and death shall have no dominion over you." The Sicilian Empedocles thought that the human soul is a mixture of air and aether, a blend of earth and heaven. He also thought that the aether "is formed by air being congealed by fire into crystalline form," as the lower air comes into contact with higher celestial fires.[9]

Others thought aether might be a kind of fire, for the sky shows bewilderingly many colors and appearances—the fiery colors of dusk, rainbows, all the varieties of clouds. It is not clear that the blue color has some special status or importance as *the* color of the sky. The Old Norse *skȳ* (whence the Middle English *skie*) meant clouds, only later including the rest of the atmosphere by association.

Thus, there are many reasons why the sky's color might not seem an important question, or why we should not try to classify it according to our common colors. There are so few color words in ancient Greek that British statesman W. E. Gladstone speculated that the Greeks were color-blind. Later scholars did not sustain this whimsical conjecture. Instead, they note that

Greek color words seem more concerned with feeling, with richness or saturation, where ours emphasize hue.[10]

Consider the specific words the Greeks used to denote blue colors. For instance, Homer uses the word *glaukos* to describe the sea and the eyes of Athena. This word might indicate the color grayish-blue, but some scholars think that it describes a bright, gleaming quality, rather than a color. It may denote the glint of light off the sea or the shining eyes of the goddess.

Indeed, the nineteenth-century art historian John Ruskin thought that Athena was "the Queen of the Air" who (among her many aspects) represented the sky itself in her blue eyes and the blue aegis or mantle she wore. According to Ruskin, her "crested and unstooping" helmet represents "the highest light of aether"; her maidenhood represents the stainless purity of the clear blue sky. Ruskin considers this sky neither distant nor merely material, for "whenever you throw your window wide open in the morning, you let in Athena, as wisdom and fresh air in the same instant; and whenever you draw a pure, long, full breath of right heaven, you take Athena into your heart, through your blood; and, with the blood, into the thoughts of your brain." In his conception, the blue sky as Athena is the source of human vitality and wisdom itself.[11]

Ruskin goes beyond the letter of Greek myths in order to touch what he thought was their unspoken spirit. His eloquent interpretations still leave us struggling with basic questions about the meaning of the simplest words, including those for color. The Greek word *kyanos* comes closest to our dark blue and is the origin of terms like cyan, cyanosis, cyanide, and Latin derivatives like cerulean. The Greeks used this word to refer to the blue of lapis lazuli, a precious stone from Egypt, Scythia, and Cyprus, now mined mostly in Afghanistan. The color *kyanos* also came to mean a dark hue, as in the blue highlights of raven-black hair.

In all these cases, the color blue refers to earthly objects and conveys a certain sense of the uncanny. Homer depicts the terrifying shield of the great chieftain Agamemnon as decorated with blue rings and snakes, bearing

at the heart a boss of bulging blue steel
and there like a crown the Gorgon's grim mask—
the burning eyes, the stark, transfixing horror—
and round her strode the shapes of Fear and Terror.[12]

Homer uses this same word *kyanos* to describe the menacing prows of warships, a dark cloud of Trojans, the blackness that engulfs the dying, or the dark brows of Zeus. Occasionally, he applies this word to rare and beautiful objects, such as the blue enamel legs of a table in the tent of Nestor, the venerable statesman.[13] In the *Odyssey*, Homer describes "a circling frieze glazed as blue as lapis" in the fabulous palace of Alcinous, king of the godlike Phaiacians.[14] Yet their blueness gives even these peaceful objects an uncanny aura. Homer may have had in mind the amazing blue friezes that adorned the Minoan palace at Knossos, the wonder of his world, site of the labyrinth and seat of ancient kings. Whether terrifying or richly rare, this intense blue stood outside the ordinary, never applied to ordinary objects. Notably, the adjective *kyanos* was not used to describe the sky.

Even later Greek authors do not discuss its color. This seems to indicate that either they had not noticed this as an unanswered question, or that the very terms "sky" and "color" did not seem compatible for them. Asking about the color of the sky might confuse earthly color with the altogether different appearances of the heavens. This is not a question of blasphemy against the superhuman powers of the sky, but of insisting on what seemed completely natural distinctions. How can we speak about the color of a feather in the same way as the color of the sky?

The early Greek thinkers tended to confine themselves to the colors of earthly objects. For instance, Plato speculated that "white and bright meeting, and falling upon a full black, become dark blue [*kyanos*], and when dark blue mixes with white, a light blue [*glaukos*] color is formed."[15] This account of sky blue as a mixture of brightness and darkness I shall call the *darkness theory*. Yet Plato noted that "God only has the knowledge and also the power which is able to combine many things into one and again resolve the one into many. But no man either is or ever will be able to accomplish either." If so, probing into the hidden nature of color is beyond human power.

But setting aside the ultimate nature of color and light still leaves the question of vision. Two opposed possibilities soon emerged. In one, the eye is active and emits rays that touch the object. In the other, the eye is a passive receptor of rays coming into it from objects. The ancient Greeks considered both seriously.

Plato favored the active view. He depicted the eye as containing a gentle fire that streams outward through the pupil and contacts the outer fire of daylight. Then these inner and outer fires coalesce to form a stream that then can contact an object, producing the sensation of sight, as if a thread of perception flows back into the eye. At night, the visual stream from the eye does not encounter a kindred fire outside and hence is quenched, cut off.[16] Similarly, Euclid and Ptolemy based their mathematical optics on the geometry of rays emerging from the eye. As late as 1280, John Pecham interpreted this to mean that these rays then return to the eyes "as messengers," rather like radar.[17] Lest all this seem too primitive, it is worth remembering that modern studies of vision have emphasized how active the eye is in visual processing, especially in its complex scanning motions. And we still speak of feeling a hard stare or looking daggers, as if they really emerged from the eye. The active quality of *seeing* is crucial to human vision, particularly in the visual arts. Indeed, the Latin word *lux* means light that is actually perceived, as opposed to the illuminating source, *lumen*. Thus, when God initially said "Let there be light (*Fiat lux*)," He was not creating a source of light (the sun was only created on the fourth day) but calling forth light as the primal act of *seeing*.[18]

The passive view began with Empedocles, who thought that all objects constantly send out effluences or little films, which then are received by the eyes.[19] This view was taken up by the early Greek atomists, who interpreted the films as atoms coming from the surfaces of bodies. Their Roman disciple, the poet Lucretius, explained vision as the reception of *eidola,* meaning "little images," which make up "a sort of outer skin perpetually peeled off the surface of objects and flying about this way and that through the air."[20]

Whether we follow the active or the passive account of vision, it is hard to understand the appearance of the sky. If vision involves a ray from the eye meeting the object, what is the object seen in the sky? At the very least, it is not like any object seen on earth. A similar difficulty plagues the passive view: from what object come the "little images" of the sky? And how does the eye then make sense of them?

Plato's student, Aristotle, continued the dialogue about these views. On one hand, Aristotle was critical of those whom he calls "the ancients." He rejected as irrational the notion that vision results from rays issuing from the eyes, for then "why should the eye not have had the power of seeing even in the dark?"[21] Clearly, he was not persuaded by Plato's attempt to have the inner fire meet the outer. On the other hand, Aristotle also rejected what he calls the "absurd" theory that colors are emanations from objects, since it neglects the role of the eye in vision.

Rather than making the eye purely active or passive, only sending out rays or only receiving them, Aristotle concentrated on what lies *between* object and eye.[22] Light itself is a state of activity, "the activity of what is transparent," of the in-between medium, when it is excited by the mutual influence of the object and the eye. Here, his word for activity is *energeia*, the vivid intercourse with the world that he calls "soul." In a powerful metaphor, he called on eating as the primal function of soul, whether in the literal chewing and digestion of food or the more subtle act of consuming involved in vision or hearing. In each case, the soul takes in something from the outside world and transforms it into itself. Mysteriously, I consume and destroy my lunch, which seems so utterly unlike my body, yet which can be incorporated into me. Similarly, through sight I take in objects outside me, somehow assimilating their alien being into *my* seeing. As I see an external object, it is now somehow *inside* me, if not literally, at least in some sense.

In both cases, Aristotle also judged that this profound transformation of the alien into myself occurs through the mediation of a transparent zone *between* the inner and outer world. He thought that seemingly empty space is really the venue in which what we call color and light emerge, as that space is more or less energized through the influence of the object seen. In this, Aristotle may have anticipated the concept of light as a *field* that emerged fully in the nineteenth century in the work of Michael Faraday and James Clerk Maxwell.[23]

Aristotle did not consider that his theory could be given mathematical form, because he judged that the unchanging forms of mathematics cannot describe the changing physical world or the dynamics of the soul. Yet he did apply his ideas to physical phenomena qualitatively. His treatise called *Meteorologica* was the beginning of meteorology, from *meteora,* meaning high, raised up, sublime, hence elevated natural phenomena. In this treatise Aristotle considered the causes of wind, rain, lightning, thunder, as well as of earthquakes, comets, and what we call meteors. He also devoted considerable attention to the rainbow, explaining it as reflection from small drops of water. These extraordinary or variable phenomena drew his attention, but the blue sky itself did not.[24]

Nevertheless, among the works traditionally attributed to Aristotle is a short treatise *On Colors,* though probably written by one of his students. In that work, the question about the sky's color seems to emerge for the first time. The author begins with the assumption that "water and air, in themselves, are by nature white," but notes also that "water and air look black when present in very deep masses."[25] He begins to connect this with other phenomena in the sky, such as its purple color at sunrise or sunset, which he attributes to a blending of feeble sunlight with thin, dusky white. In general,

> we never see a color in absolute purity: it is always blended, if not with another color, then with rays of light or with shadows, and so it assumes a tint other than its own. . . . Thus all hues represent a threefold mixture of light, a translucent medium (e.g., water or air), and underlying colors

from which the light is reflected. . . . Air seen close at hand appears to have no color, for it is so rare that it yields and gives passage to the denser rays of light, which thus shine through it; but when seen in a deep mass it looks practically dark blue.[26]

Thus, the sky is blue because it lets through the surrounding darkness, a notion close to Plato's idea that I have called the darkness theory. Aristotle's student finds confirmation in the deepening blue of the sky at nightfall, "for where light fails, the air lets darkness through and looks dark blue," though air, by itself, is "the whitest of things."

Though much about this explanation will later be criticized and revised, it makes several crucial steps. First, this is the earliest text that recognizes that there *is* a question, that the blueness of the sky needs explanation. This is no small matter, for an unasked question finds no answer. Second, this account attributes the color of the sky to the interaction between the air and outside influences, as opposed to attributing the color to bodies floating in the air, but not the air itself. This powerful assertion will remain in doubt for almost two thousand years.

Aristotle himself would probably have found his student's explanation puzzling, if not contradictory. Aristotle had argued that darkness was not the *presence* of something, but the *absence* of light. This follows from his basic concept that light is a state of energized activity of a medium, so that darkness is the lack of that activity. How, then, can the outer darkness shine through the atmosphere? Also, the air seems passive, almost irrelevant, in this explanation, which puts the onus on a paradoxically potent darkness.

In this way, Aristotle and his student came close to identifying the sky's blue with the interaction of light with air. Their hesitation is understandable, for if air is truly transparent, it is hard to understand why it turns *blue*, rather than some other color. Hence, they needed the darkness of the night sky to mix with the white of the air. Indeed, Aristotle himself had speculated that what we call the different colors are really formed by various mixtures of minute quantities of black and white, perhaps arranged in simple ratios like 3:2, such as characterize the consonant intervals of music. This ingenious theory goes below the threshold of visible size to provide what we would call a microscopic basis for color, one that would explain the properties of the light-bearing medium. In hindsight, his musical theory of black and white building blocks may seem a kind of atomic theory.[27]

However, Aristotle definitely rejected the idea of atoms flying around in the void, which particularly troubled him. How can *nothingness* exist without paradox and contradiction? How can nonexistence exist? Instead, he argued that the cosmos is continuous and cannot have any void spaces. If so, there are no fundamental microscopic structures whose size or properties might explain the sky's blueness.

The surviving writings of the earliest Greek atomists are too fragmentary for us to know whether they puzzled over this question or had developed their theory enough to address it. The earliest of them, Leucippus, wrote two centuries before Aristotle and left only one fragment: "Nothing happens at random; whatever comes about is by rational necessity," presumably because of atoms in motion.[28] If so, surely the appearance of the sky also comes about through atoms. Lucretius sang of atomic theory as a remedy for human fear and superstition. He took particular care to show that "the sky in all its zones is mortal," demystifying the realm of Zeus and dissolving his thunderbolts into atoms. Lucretius is one of the few ancient authors who refer explicitly to "the blue expanses of heaven" as he explains different clouds through their underlying atoms.[29]

Yet Lucretius did not address the cause of the sky's color (though he wondered why thunder could come out of a blue sky) and his explanations of thunder and lightning as clashing atoms remained quite schematic. To gain explanatory power, the atomic hypothesis required deep transformation. Even so, it proved to be crucial.

2 Ultramarine

After these suggestive beginnings, the quest did not advance for a thousand years. Aristotle remained the preeminent source for further thought. Muslim thinkers carried his astronomical theories further but also inherited the presumption of seven celestial spheres surrounding the earth. Over time, these spheres became increasingly solid and substantial. The astronomer Ptolemy introduced them as purely mathematical contrivances, but later treated them as physically real. The more substantial they became, the more the question of the sky's color was not a matter of the earth's atmosphere but of the spheres outside.[1]

Gradually, however, attention turned to the nature of the atmosphere and its relation to the light of the sky. This inquiry was considerably more advanced in China than in Europe. Already in 400 A.D., Jiang Ji offered as an explanation of "why the sun appears red in the morning and evening, while it looks white at midday" that "the terrestrial vapors [*di qi*] do not go up very high into the sky," for if they did, the sun would always look red. Thus, Jiang Ji already recognized that the sun at the horizon is seen through a thicker section of the earth's atmosphere. A century earlier, other Chinese scholars attributed misty skies to small suspended particles.[2]

As in so many other areas, here too Arabic thinkers were far in advance of the West. Their activity was supported by *ḥadīth*, sayings attributed to the Prophet Muḥammad himself: "He who pursues the road of knowledge God will direct to the road of Paradise. . . . The scholar's ink is holier than the martyr's blood. . . . Seeking knowledge is required of every Muslim."[3] Among Muslim philosophers, Ya'qūb Ibn Ishāq al-Kindī was among the greatest. Active in Baghdad in the ninth century, al-Kindī was a luminary of the House of Wisdom, an extraordinary center of study and scholarship founded by a son of the caliph Harun al-Rashid, he of the Thousand and One Nights.

Al-Kindī was one of the first to sponsor translations of Aristotle. His own interests spanned the natural sciences, mathematics, optics, music, and cosmology—he was a true polymath long before the Renaissance in the West. His concern with the operations of the physical world influenced Islamic philosophy; his devotion to reason rendered him suspect in the eyes of the ultra-orthodox.

Al-Kindī turned away from Aristotle and returned to the active concept of vision as radiation emanating from the eye. Consider his letter "regarding the cause of the azure-blue color that is seen in the atmosphere in the direction of the sky and that is considered to be the color of the sky."[4] That is, azure is not the color of the sky itself, but "simply a thing that appears to the sight," a product of human vision. He argues that only solid objects are fully lighted and hence really have a color, whereas bodies that are not solid like water, air, fire, and the transparent sky do not get completely lighted and hence have no color.

According to al-Kindī, "the air surrounding the earth gets weakly lighted by the earthly particles dissolved in it and changed into fiery ones due to the heat that they have accepted from the reflection of the rays of the earth" and of starlight. These lights "intermingle into a color in the middle of shadow and light and this is the azure-blue color." Thus, his explanation rests on the "earthly particles" suspended in the atmosphere that scatter light from the sky as well as that reflected from the earth. Let us call this the *dust theory*.

This theory represents a notable shift from the darkness theory of Plato and Aristotle. For the first time, al-Kindī gives an explanation that points to something in the atmosphere itself scattering light, though without fully clarifying what these "earthly particles" might be. Most obviously, they could be dust, omnipresent in the Middle Eastern deserts, but one wonders whether al-Kindī worried about what would happen when the airborne dust settles.

In optics, al-Kindī's tenth-century successor Ibn al-Haytham, known in the West as Alhazen, went even further.[5] Ibn al-Haytham came to the attention of the caliph of Egypt for his claim that he could erect constructions that would regulate the flow of the Nile. But when Ibn al-Haytham investigated further, he was dismayed to discover that the ancients who had built the magnificent monuments of ancient Egypt had not been able to achieve what he proposed. Accordingly, he gave up his plans. To avoid the murderous anger of the caliph, Ibn al-Haytham then feigned madness for some years, until the caliph died and he could resume his life as a sane man.

Ibn al-Haytham's *Optics* was the most impressive work in natural philosophy since the death of Archimedes. In it, Ibn al-Haytham treated in great detail reflection (the return of light rays impinging on another medium, as from air on a mirror) and refraction (the bending of light rays penetrating obliquely from one medium to another, as from air to water). Departing from al-Kindī's view, Ibn al-Haytham considered vision to be rays emanating from the object that pass into the eye, whose anatomy he studied carefully. He did not need *eidola* or emanations, only rays that reflect from object to eye following Euclidean geometry. His contemporary, Ibn Sīnā (known in the West as Avicenna), considered the rays to be composed of small particles; both of them believed that the rays' velocity was very great but finite (a view that remained dormant in the West until Ole Roemer's measurement of the velocity of light in 1676). Ibn al-Haytham pursued experimental verification with special care, designing and executing his experiments with the kind of attention known in the West only after 1600, when experimentation became the touchstone of the "new philosophy" pioneered by Galileo.

For instance, to test the rectilinear propagation of light rays (assumed by Euclid and Ptolemy in their optics), Ibn al-Haytham carefully drilled holes in two walls, some perpendicular, others oblique but converging on the same source. He tested direct and reflected light, colored light, and the red light of dawn. He studied in detail the properties of curved mirrors and lenses, unifying the mathematical approach of Euclid with the concept of rays streaming into the eye. Ibn al-Haytham's studies yield an understanding of the basic anatomy of the eye that is very close to our modern understanding. He projected the image of a solar eclipse through a small hole in a wall, a device known in the West much later as a *camera obscura*. He also determined (correctly) that the last glimmer of twilight occurs when the sun is fully 19° below the horizon.

Ibn al-Haytham argued that the sea's color only reflects that of the sky. He noted that the blue of the zenith is "purer and nearer to darkness than that which is toward the horizon." This distinct whitening of the horizon compared to the zenith can be noted on any clear day, though no one had recorded it before him. He went further to derive this from the fact that "the distance of the length of rays from the particles of the atmospherical sphere [the aether surrounding the earth] is different to the sight; that which is towards the zenith is shorter and that is towards the horizon is greater." Here, Ibn al-Haytham made important use of his basic hypothesis that light is rays whose appearance depends on the length they travel through the atmosphere. "If there are solid substances in the air like the vapor and the smoke and the light flashes upon them, there appear colors like the colors of the clouds and so on, according to their grades as regards thickness, thinness, clearness, and turbidness."

Like al-Kindī, Ibn al-Haytham held that the light in the sky is "reflected" from "solid substances" suspended in the air. But instead of dust, Ibn al-Haytham considers "vapor" or smoke, which we will refer to as the *particle theory*. This important generalization may already have occurred to the Chinese (though they may not have thought of dust particles as solid). Ibn al-Haytham probably knew nothing of these Chinese thinkers, but, like them, he noted that sunset reddening depends on the greater distance horizontal rays must travel through the atmosphere, compared to those coming from the zenith, once again confirming the utility of his basic hypothesis about light rays. Likewise, he noted the "somber color" when dust hangs in the air.

But Ibn al-Haytham also alludes to the obvious possibility that rain might clear the air of dust—and how then could the blue color remain? Do "solid substances" like vapor or smoke never settle, behaving essentially unlike dust? Even if we assume that particles always manage to hang in the air, it is not clear why they make the sky *blue*, rather than some other color. Ibn al-Haytham's discussion ends inconclusively, as if he too were aware of the importance of these unsolved problems. His final words show his perplexity: "God knows what is right."

There the matter rested, for his immediate successors went no further. His younger contemporary al-Bīrūnī traveled to Mount Demavend (18,600 feet or 5,600 meters), the highest mountain in Persia. It is not clear how high he climbed, but al-Bīrūnī went high enough to note that "since the mountain is so high, we have indeed witnessed [the blackening of the air] . . . at its summit." Presumably, he was referring to the deeper blue seen at high altitudes; if so, he was apparently the first to record it. "Blackening of the air" captures the extraordinary darkness of the sky at such heights, especially viewed against bright snow. But he may also have been echoing the ancient conception that the sky is a veil over outer darkness. Thus, he may well have thought he was not observing the color of the sky but rather the blackness that surrounds it.[6]

His remarks come in his "Exhaustive Treatise on Shadows," whose name significantly anticipates the deeper shadow that fell on Islamic science. In 1258, the Mongols invaded Baghdad, overturned the 'Abbasid caliphate, and laid waste its celebrated House of Wisdom. But its great achievements had fallen into oblivion even before the barbarians arrived. In late-thirteenth-century Cairo, the Muslim jurist al-Qarafi responded to the question why the sky was blue by echoing the ancient darkness theory that it is "a mixture of the black with the clear." After two centuries Ibn al-Haytham's great *Optics* had been forgotten by his own successors.[7]

Though forgotten in the East, this work reappeared in the West not long after Aristotle's works were translated into Latin in the twelfth century. A thirteenth-century Pole named Witelo repeated Ibn al-Haytham's arguments without acknowledging their source; in the sixteenth century, a full translation of Ibn al-Haytham's *Optics* appeared.[8] His work began to live again, notably in the writings of the thirteenth-century scholar Roger Bacon, a controversial figure who struggled to escape ecclesiastical condemnation. Bacon's *Opus Majus* (*Great Work*) is an impressive summation of his vision of a revived natural science, strongly reliant on mathematics and experiment. The optical part of this work relies on Ibn al-Haytham, though it generally does not go much beyond him.

But consider Bacon's comparison of the sky to a body of water. "For the parts of deep water cast forward a shadow on those that succeed them, and a darkness is produced that absorbs the quality of the rarity, so that in this way the whole body of water appears like some dense body, and the same is true of the air or celestial transparent medium at a distance, for which reason it is rendered visible, but not so at close range."[9] The greater the depth of water or air, the deeper the blue.

Bacon's analogy emphasizes the darkening blue in a continuous medium. Though he did not draw this conclusion, we are tempted to consider that the blue of the sky might similarly come from the direct interaction of light and atmosphere, without the necessity of solid particles. Yet this speculation is far from what Bacon had in mind. He goes on to say that in deep water "color appears owing to shadows projected by particles." Contrary to our suppositions, Bacon thought of water as containing "particles," meaning "parts," not atoms in our sense, and likewise for air. His emphasis on the "shadow" cast by these "parts" shows his real thrust: the blue is a kind of shadow. What is more, Bacon goes on to note that, according to theologians, there is a "heaven of water" beyond the stars, necessitated by the verse in Genesis about the "waters above the firmament." Suddenly we realize that his comparison with water is no mere analogy: he really thought that beyond the sky *was* a body of water. Here, the intrusion of biblical theology undermines an interesting line of thought. Yet by invoking "parts of air" rather than dust, Bacon opened the possibility that the atmosphere itself, rather than any suspended substances, might be the source of the blue. More likely, however, he was merely rephrasing Ibn al-Haytham's particle theory.

Bacon lived during a curious shift in color preferences. During the eleventh and twelfth centuries, blue began to acquire a new kind of symbolic and aesthetic value it had previously lacked. Gold, not blue, was used for the sky in Byzantine and early Italian painting. A literal match of colors was less important than the ethereal ideality of gold as emblematic of heavenly light. In the fourth century, Saint Gregory of Nyssa identified two kinds of divine apparition, one through light and the other through darkness, as when Moses encountered God first in the burning bush and later in the darkness. Two centuries later, the writer known as Pseudo-Dionysius asserted that "the Divine Darkness is the inaccessible light in which God is said to dwell." Hence, he considered the dark horses of the Apocalypse to be blue because of the "hidden depths of their nature."[10]

In many Jewish, Neoplatonic, and Christian sources, God is identified with space and space with light, but in the early Middle Ages blue was considered akin to darkness. For instance, a mosaic panel in San Apollinare Nuovo in Ravenna depicting the Last Judgment shows a red angel to Christ's right, with the

Newton's theory did not do justice to the true nature of light.

When Goethe picked up a prism and held it before his eyes, he was startled to see not Newton's spectrum but rather the ordinary images of things, fringed with colors. Goethe took this as evidence that "Nature falls silent on the rack. Her true answer to a sincere question is: Yes! yes! No! no! Anything more is bad." Goethe concluded that Newton was wrong and wanted to liberate "the phenomena once and for all from the gloomy empirical-mechanical-dogmatic torture chamber."[7] As he put it in a poem satirizing Newton's "narrow mind,"

Friends, flee the dark chamber,
Where the light is entangled
And in most wretched distress
Stoops to perverse images.

Outside, gazing at "the heavenly blue of the serene day," the poet glorifies nature, "happy, sound in eye and heart, / and recognizes the universal, eternal principles of color."[8] Unfortunately, his views led Goethe to reject the discovery of discrete lines in the solar spectrum because he felt the light had been "tortured" by passing through a narrow slit.

Goethe also reacted strongly to a controversy in the 1820s about whether the sky's color was an optical illusion based on complementary colors. He repeated the simple experiment involved (see experiment 5.1 in appendix A) and was incensed to think that the reality of this most basic phenomenon might be questioned. In response, Goethe formulated his own color theory.[9]

In his theory, Goethe made much of Saussure's observations of the colors of shadows cast at different altitudes. Shadows are not absolutely black but blue, especially seen against the high contrast of snow. Saussure had noted that shadows are less blue at high altitudes, more blue at lower, and he took this to confirm the view that "these colors arise from accidental vapors diffused in the air, which communicate their own hues to the shadows."[10] Goethe agreed and interpreted the lack of blue shadow at high altitudes to indicate that the incident sunlight is colorless, and hence the blue-colored shadows at lower altitudes are caused by "vapors," by which he and Saussure apparently meant small particles. Modern usage restricts the word vapor to mean only a gas, so that "water vapor" strictly refers only to the gaseous form of water, not to water droplets. As we will see, this important distinction emerged only slowly.

Goethe felt that Newton in his dark chamber neglected these outdoor phenomena. Instead, Goethe studied the practice of painters, especially as discussed in Leonardo's *Treatise on Painting*. Goethe tells the story of a painter who was trying to restore a portrait of a man dressed in black and began by wiping the canvas with a wet sponge. To his dismay, the black robe suddenly turned blue. Goethe notes correctly that the varnish had become turbid and thus looked blue, in the way smoke does. The next day, after the water had evaporated, the blue color vanished.[11]

Goethe interprets this as one example of what he calls the urphenomenon (*Urphänomenon*), the primordial or archetypal phenomenon, namely light seen through a turbid (not perfectly transparent) medium. "A turbid glass held before a dark background and illuminated from the front will appear bluish. The less turbid the glass, the bluer it will look; the least turbid glass will seem violet. Conversely, the same glass held before something bright will look yellow. The denser the glass, the redder it will seem, so that in the end even the sun will appear ruby red."[12] In modern terms, he was observing the difference between colors seen by *scattered* and *transmitted* light. Goethe took this as not merely an isolated experiment but a touchstone, the key to the whole nature of color. He found many other examples of it, including dust suspensions, the lower part of a candle flame, and clear liqueur (see experiment 5.2 in appendix A).

Goethe's scientific approach relied on identifying archetypes, which he used as central organizing insights. In his studies of plants, he searched for what he called the *Urpflanze,* the archetypal plant, of whose anatomy all known plants are variants.[13] Likewise, the *Urtier* would be the archetypal animal. In such studies, Goethe took part in early phases of what we now think of as morphology, the study of the commonalities and differences in biological structure that later became an important element in Darwin's evolutionary synthesis. In linguistics as pioneered by Alexander von Humboldt's brother Wilhelm, the search for an *Ursprache,* a primordial or ancestral language, led back from the known European tongues to their common ancestor, Indo-European.

The ultimate expression of Goethe's urphenomenon is the blue sky itself. "The air, even at its clearest, is a vehicle for moisture and must therefore be considered a turbid medium. This is why the sky opposite the sun and around it looks blue: the darkness of space creates this effect through the veiling. This is also why mountains in the middle distance seem darker blue than those in the far distance," though here again Goethe ignores the difference between gaseous water vapor and condensed droplets. He also contradicts Leonardo's dictum that more distant mountains are bluer:

On the highest mountain peaks the air will seem deep blue because of the purity of the atmosphere there; ultimately it will take on a reddish tinge. In the plains, where the air becomes increasingly dense and filled with turbidity, the blue will grow ever paler, finally vanishing and assuming a completely white appearance. Seen through an atmosphere thick with haze, the sun and the bright area around it will seem to have a yellow-red to red color. . . .[14]

Thus, the sun would appear reddish, even more so at sunset. In a way, Goethe's views synthesize the dust theory and the darkness theory. For him, the urphenomenon expresses a pervasive motif in nature, rather than its causes: "Let us not seek for something behind the phenomena—they themselves are the theory."[15]

Goethe's arguments about colored shadows points toward the complex physiology and psychology of color vision, which Newton had tried to separate from the physics of color. Goethe sought to unite human perception with objective nature, in the spirit of *Naturphilosophie*. When the celebrated chemist Count Rumford called colored shadows optical illusions, Goethe answered heatedly that "optical illusion is optical truth. It is blasphemy to say there is such a thing as an *optical illusion*."[16] What we perceive should not be explained away or treated differently from "real" phenomena.

Thus, Goethe gave expression to the felt perception of color with its human and artistic associations, which go beyond pure physics. Though his theory was generally rejected by physicists, it directly influenced the founders of modern physiological optics. Goethe's work may thus be seen as leading to this new science of human vision, which enforced the separation between physiology and physics that he so resisted. Goethe's color theory also had a deep and abiding influence on philosophers and artists, who recognized their feelings in his descriptions.[17]

For instance, Goethe classified blue on the "*minus*" side of colors, producing "a restless, susceptible, anxious impression." In contrast, "*plus*" colors like yellow "are lively, quick, aspiring." Blue, then, has "a peculiar and almost indescribable effect on the eye. As a hue it is powerful, but it is on the negative side, and in its highest purity is, as it were, a stimulating negation. Its appearance, then, is a kind of contradiction between excitement and repose." This paradox expresses the romantic ambiguity of blue, its fascination and remoteness. First, it seems removed, for "as the upper sky and distant mountains appear blue, so a blue surface seems to retire from us. But as we readily follow an agreeable object that flies from us, so we love to contemplate blue, not because it advances to us, but because it draws us after it." Here, as in Novalis, blue is given distinctly female associations, recalling Goethe's *Ewig-Weibliche*, the "eternal feminine" that "draws us on" at the conclusion of his *Faust.*[18]

Goethe judged that, though blue fascinates, it finally "gives us an impression of cold, and thus, again, reminds us of shade." Its affinity with darkness harks back to Plato; its coldness recalls its ancient association with death. Rooms hung with blue seem large, but "empty and cold"; "the appearance of objects seen through a blue glass is gloomy and melancholy." These associations resonate with the romantic sensibility, drawn toward a sublime object that annihilates common feeling in the shudder of infinity. Though one wonders how far such color associations are culturally conditioned, these remain commonplaces of modern design and color theory: blue is a "cold" color, which seems to recede, compared with a "warm" color like red. It remains to be seen whether the romantic attitudes toward color are a passing fashion or a deep insight into human sensibility.[19]

Though they parted ways, physics and *Naturphilosophie* did interact. In 1801, Johann Ritter, a student of *Naturphilosophie,* discovered ultraviolet radiation by observing that the solar spectrum seemed to act on light-sensitive emulsions even past the violet. Hans Christian Ørsted connected his discovery of electromagnetism (1820) with the search for a "unity of all the forces" of nature, as Humboldt had envisioned, a search that still engages modern physicists in their own way. From the vantage point of *Naturphilosophie,* the coming of the wave theory of light rebuked the materialism and mechanism of Newtonian physics. Where the particle is discrete, separate, intensely apart, the wave is collective, mutual, united.[20]

Though some, like Euler, had advocated wave theory, its experimental vindication long hung in the balance after Grimaldi's initial work in the 1660s. He had been using rather large beams of light, about a centimeter wide; he discovered diffraction only when he narrowed the beam down to an extremely narrow pencil of light emerging from a pointlike source into a totally darkened room. If Grimaldi had used more of those narrow beams, he might have anticipated by more than a century the crucial confirmation of the wave theory that Thomas Young gave in 1801. Though a practicing physician (and an authority on Egyptian hieroglyphs), Young also devoted himself to experimentation on the nature of light. He repeated Grimaldi's observations by making light pass over a narrow card or thread (see figure 3.5), noting the pattern of fringes. But

Young felt that this did not yet give definitive proof that light interferes with itself, in the way that water or sound waves do.

To investigate this, Young illuminated *both* edges of a thin slip of card with strong light and noted on the wall a pattern of bright and dark fringes he called "interference" (figure 5.1; see experiment 5.3 in appendix A). Young argued that "these fringes were the joint effects of the portions of light passing on each side of the slip of card, and inflected, or rather diffracted, into the shadow."[21]

To prove this, Young held a little screen so that it would block the light from reaching one edge of the card, so that only *one* edge was illuminated. Then "all the fringes which had before been observed in the shadow on the wall immediately disappeared," though the incident light remained unchanged and the one edge was still illuminated as before. This proved that "one of the two portions [of the card] was incapable of producing the fringes alone." Thus, the fringes showed that *both* edges were needed and thus that the light had to illuminate *both* of them at once, something particles could not do, but waves do naturally (see experiment 5.4 in appendix A).

Figure 5.1
The overlapped light from two slits showing the fringes of interference, showing the result of Thomas Young's experiment.

Newton's test had been met. Light indeed "bends around obstacles" as he had demanded, though only slightly if the obstacle is as large as a thread, indicating that the light's wavelength must be very small. In his experiments on "Newton's rings" produced by the pressed glass plates, Newton had made measurements of the frequency of vibration of his "fits," finding extraordinarily large numbers, which are proportional to the inverse of the wavelength involved. When this is taken into account, Newton's results tally with Young's measurements of wavelengths.

As a physician, Young had investigated the physiology of the eye and suggested that only *three* color receptors in the retina (red, yellow, and blue) would suffice to give full color vision, rather than "an infinite number of particles, each capable of vibrating in perfect unison with every possible undulation," as Newton thought.[22] Young also studied the physiology of the ear and the wave nature of sound. His joint study of vision and hearing helped him see how the wave theory could encompass both. For instance, in his experiment with the two pinholes, he was able to use Christiaan Huygens's pioneering ideas about waves to explain the interference he observed. In 1690, Huygens had stated a general principle: At every point, we can visualize an incoming wave generating a new wave, so that all these "secondary waves" join together to form the wave as it propagates (figure 5.2). To be sure, this mathematical construction is unphysical; Huygens imagines waves regenerating even in empty space, though physically this can happen only where matter can absorb and re-radiate the waves. Yet Huygens's principle gives a faithful *virtual* representation of the way waves propagate. Indeed, it seems astonishing that the image of a flame reaches the eye so faithfully, if indeed its waves

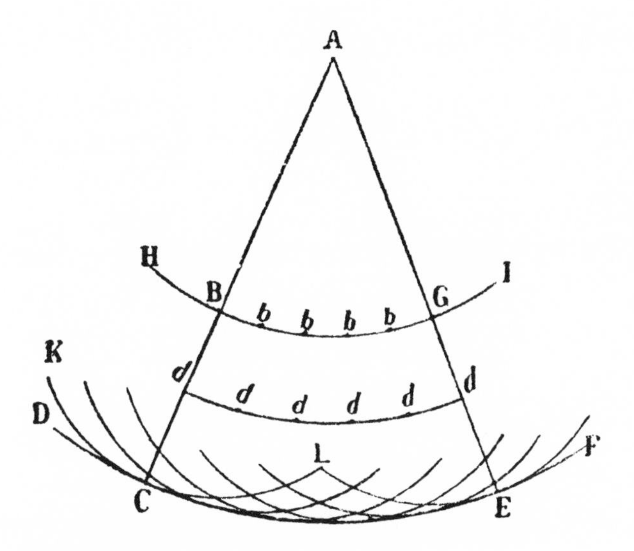

Figure 5.2
Huygens's diagram of a wave emerging from *A* and generating secondary waves at all points *b* and *d*, forming the wave front along *DCEF* (*Tractatus de Lumine*, 1691).

regenerate innumerable times as they travel. Huygens answered that "we will cease to be astonished when considering that at a great distance from the luminous body an infinity of waves, even if emitted by various parts of this body, join together forming one single wave which consequently is strong enough to be detected." The integrity of the flame's image is maintained by the mutual interference of all its waves (figure 5.3).[23]

After all, as waves travel, only their form truly advances, not the medium; the buoy moves up and down, but only the wave moves forward. Unlike particles, a wave is not a substance, but a process that continually regenerates itself; as

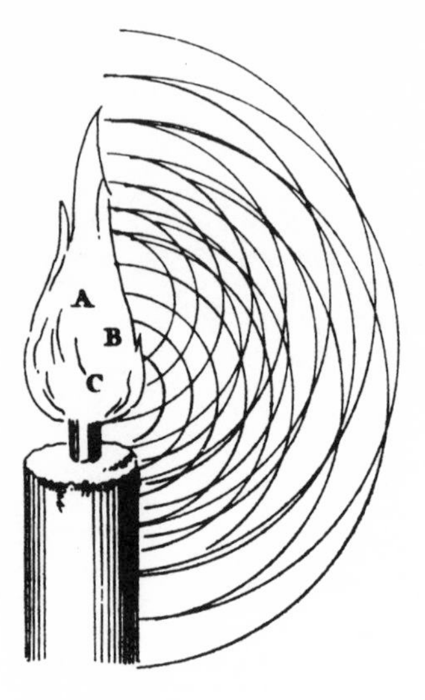

Figure 5.3
Huygens's depiction of different points in a flame emitting waves (*Tractatus de Lumine*).

waves are absorbed in their passage, they are then re-radiated, ever dying and reborn. Shakespeare already saw it on the beach:

> Like as the waves make towards the pebbled shore
> So do our minutes hasten to their end;
> Each changing place with that which goes before,
> In sequent toil all forwards do contend.[24]

Young applied Huygens's principle to his experiment with two pinholes to show how naturally the wave theory would explain the pattern of fringes that he observed. Though he used the term "interference," Young's explanation also applies to the phenomena much earlier labeled "diffraction," so that (despite attempts to distinguish them), in essence diffraction *is* interference. Both are manifestations of the same physics: the scattering of light by matter.

The wave theory of light represented such a challenge to the dominant Newtonian theory that many years passed between Young's announcement of the interference of light and the full recognition that the wave theory could readily explain these new phenomena. Even in the 1850s many still defended the older Newtonian view.[25] But the French engineer Augustin

Figure 5.4
Augustin Fresnel.

Fresnel (figure 5.4) came to the same conclusions as Young and was able to give a complete and satisfying mathematical account of diffraction and interference. Consider, for example, his beautiful diagram of the interference of waves passing around an obstacle (figure 5.5). Siméon-Denis Poisson noted that Fresnel's theory implied that, at the very center of the shadow cast by an opaque, circular screen, there should be a minute spot of light. Close observation then confirmed this striking prediction, apparently inexplicable by the particle theory (how could a spot of light appear at the center of a dark shadow?) but understandable in waves that meet and reinforce each other at the center (figure 5.6). Arago proposed another crucial experiment: by measuring the speed of light in water, in 1850 Jean Foucault determined that it was less than in air, not greater, as Newton had argued.[26]

As the mathematical theory of waves advanced, it also confronted the problem of the sky's color as a crucial test. But before that could be consummated, an entirely new aspect of light emerged that would radically alter the problem itself.

This story began in 1669 with the discovery of a peculiar property of the crystal of calcite, readily available at most rock and mineral shops. Erasmus Bartholinus, a Danish physician, noticed that, if you look through such a crystal at a piece of paper with a dot on it, you see two dots, not one. If you rotate the crystal, one dot seems to move with it, while the other stays still. Bartholinus called the moving ray the "extraordinary" image, and the stationary one the "ordinary" image. This "double refraction" (as he called it) seems mysterious; how can one ray of light split into two?[27]

Then in 1690 Huygens investigated further by putting two of these crystals in series. The first divides the initial image into

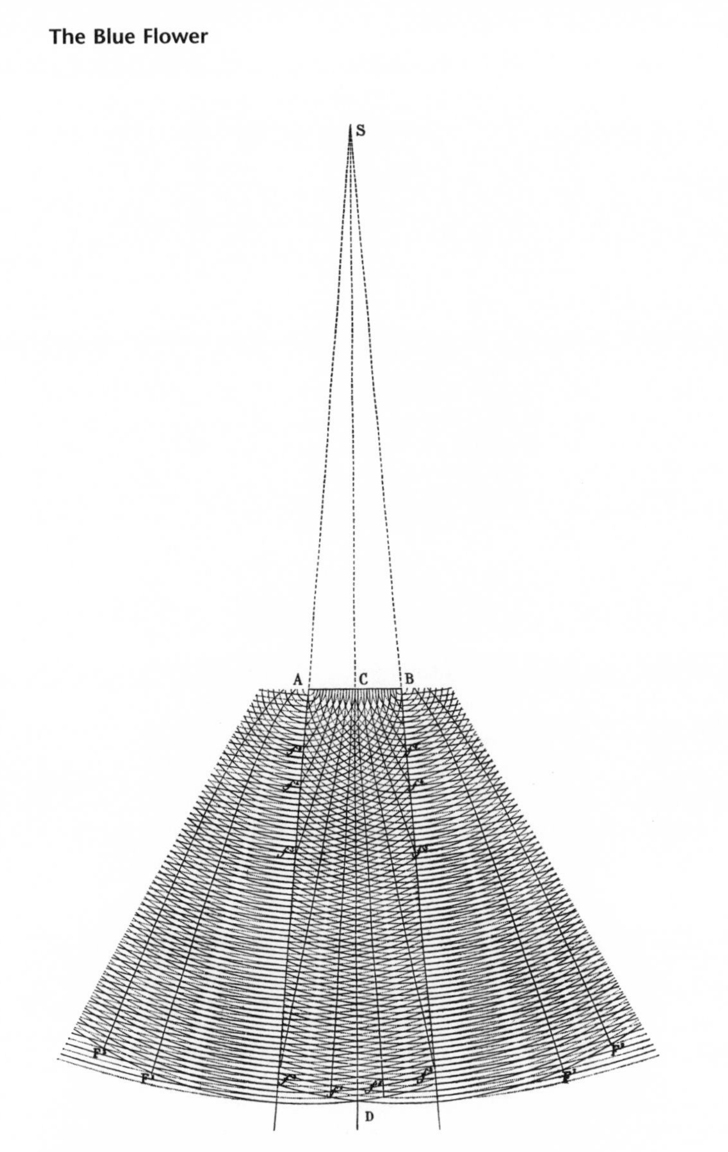

Figure 5.5
Fresnel's diagram showing light emerging from a source *S*, meeting an opaque obstacle *ACB*, then generating a diffraction pattern.

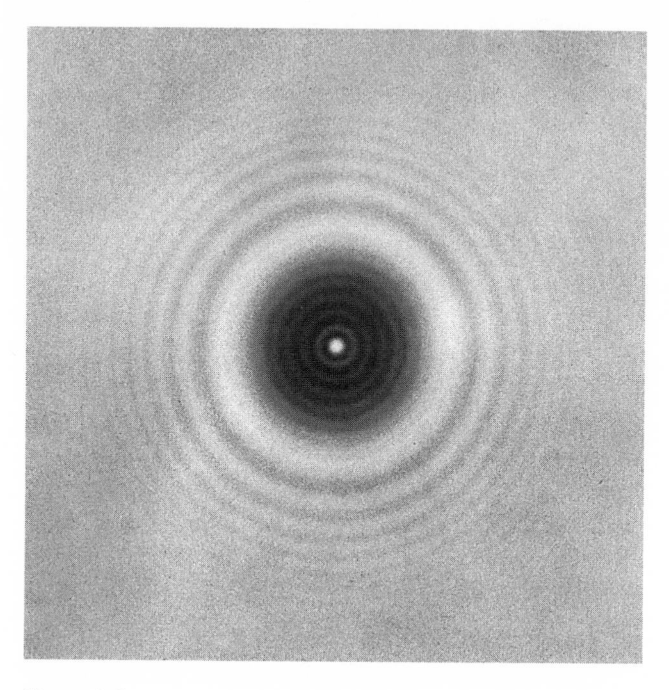

Figure 5.6
The shadow of an opaque, circular screen, showing the bright "Poisson spot" in the center predicted by Fresnel's theory.

two, and in many cases the second doubles these again, making four images of the original dot. Yet if the crystals are turned in certain orientations with respect to each other, the second crystal allows the two rays to pass through it *without* doubling. Huygens, already a believer in the wave theory of light, concluded that "the waves of light, after having passed through the first crystal, acquire a certain form or disposition" that allows them to pass through the second without being divided yet again. He had no explanation for this strange "disposition."[28]

Hooke had speculated that the waves of light were "transverse," meaning that the vibrations were not along the line of motion of the wave (as with sound waves) but at right angles to the motion of the wave (as with water waves). Hooke's suggestion was confirmed by Young only in 1817.[29]

Meanwhile, in 1808 Étienne-Louise Malus looked through a calcite crystal at the light reflecting from a window in the Luxembourg Palace opposite him. As he rotated the crystal, he noticed that the two images he saw through the calcite were alternately extinguished. To describe the behavior of the light, Malus used the term *polarization,* as we still do today.[30] But today, we don't need a crystal to get polarized light; we need merely look at light reflected from some surface.

In 1812, David Brewster, one of the last loyalists to Newton's particle theory of light, gave a simple mathematical law for this polarization by reflection: Light incident at a certain angle, "Brewster's angle," is completely polarized, whereas light incident at other angles is only partially polarized.[31] For example, sunlight reflecting from ground and cars becomes partially horizontally polarized; thus, your polarized sunglasses, aligned to admit only vertically polarized light, will block the horizontally polarized glare. (Try looking through polarized sunglasses as you rotate them; the glare increases.) In contrast, light coming from an incandescent bulb is *unpolarized* (rotating your sunglasses makes no difference).

The first decades of the nineteenth century brought forth a rush of further discoveries that completed the picture. Arago discovered in 1811 that polarized light passing through certain materials (such as a solution of D-glucose (dextrose) dissolved in water) will remain polarized, but its direction of polarization will change, rotating with respect to its original direction. In that same year, he showed that the light in the sky itself was polarized in a direction at right angles to the sun. There are also certain "neutral points" in the sky at which the polarization vanishes; Brewster noted that they changed with the sun's altitude. These are crucial developments: No longer an "extraordinary" phenomenon, polarization was ubiquitous. Also, it took center stage in the study of the light from the sky, for any explanation would have to come to terms with the polarization of that light, not just its color.[32]

At first, these polarization effects could be observed only by using crystals. But in 1844 Wilhelm Haidinger, an Austrian minerologist, discovered evidence of polarization of sky light that is visible to the naked eye. Looking into a clear blue sky, a trained observer can discern a faint pattern called "Haidinger's brush," about four degrees wide, symmetric, bow-tie-shaped, and yellowish (figure 5.7; see experiment 5.5 in appendix A). Thus, the eye has some receptors in the macular area of the

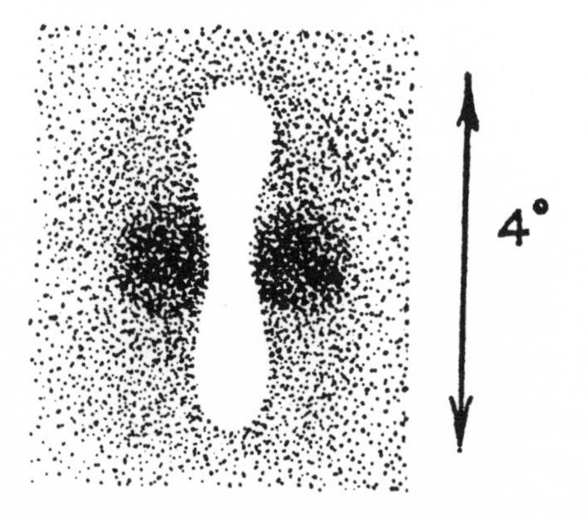

Figure 5.7
Haidinger's brush, when the polarization of the light is vertical.

retina that are sensitive to the polarization of the incident light. Why blue and yellow, rather than other colors, are seen in the brush is still an unsolved problem of the human visual system.[33]

Even without such evidence as Haidinger's brush, though, the phenomenon of polarization seemed to deal the particle theory of light another crushing blow. If light is particles, what about them could possibly correspond to polarization? But the wave theory had a simple explanation. If, as Hooke and Young argued, these were transverse waves—vibrating at right angles to the direction of motion of the wave—this would lead to the possibility of horizontally or vertically polarized light.

This helped explain another discovery of Fresnel's. He duplicated Young's experiment of the two slits, but with the light coming through one slit polarized horizontally, and the other polarized vertically. In that case, there was no interference, which puzzled Fresnel. But Young pointed out to him that light waves can interfere only if they are polarized in the *same* direction so that they can reinforce or cancel each other in that direction. Hence light waves polarized at right angles cannot interfere.

In contrast, water waves cannot be polarized, though they are transverse, because they wave in only *one* dimension, up and down. But light waves have *two* dimensions available in the transverse plane, which allows them to be polarized. Accordingly, we must be wary of analogies that confuse these fundamentally different kinds of wave. The problem then becomes how to apply this new conception of light as a wave to the problem of the light in the sky. How does light interact with matter? How does that interaction result in a blue color?

Arago revived the cyanometer in 1815, making a more sophisticated version (figure 5.8) to measure the relation between the

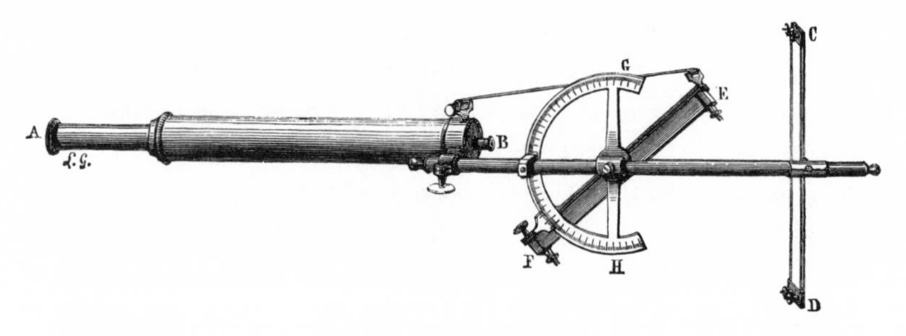

Figure 5.8
Arago's cyanometer (1815). The mirror *F* can reflect to the eye *A* the light from different parts of the sky, which can then be seen against a color scale, *CD*.

polarization and color of the sky's light.[34] To that end, he also developed the polarimeter, a device using a succession of crystals and reflecting plates to measure to what degree the light was polarized and to what degree unpolarized. With this instrument, he could tell whether light was reflected from solid or liquid surfaces or emitted from incandescent gases, because of the different degrees of polarization characteristic of those states. This was an important tool for astronomy that enabled him to determine whether the object he saw in his telescope was solid or gaseous, however distant it might be. Further, knowing what happens to polarized light upon reflection, he could deduce whether the light he was seeing had been reflected or not. In this way, he observed that the edge of the sun is gaseous and that the rays from the sun are refracted, not reflected, in the atmosphere of the earth. This implies that the polarization of the sky's light is not due to mirror-like reflection from particles in the atmosphere, as Bouguer had hypothesized.

So far, this seems to confirm the dominant opinion that the sky's color is due to refraction of the sun's light by droplets in the atmosphere. Even so, some continued to hold the view that air is "a coloured fluid" that is "naturally blue," as the natural philosopher John Leslie asserted in 1838.[35] In 1840, James David Forbes noticed that sunlight passing through just-condensing clouds of steam looks red. He argued that this would account for the red color of sunsets, but he did not wonder what happened to the blue wavelengths of incident sunlight, which might have led him to anticipate some crucial discoveries to come.[36]

Rudolf Clausius, the great pioneer of thermodynamics, also tried his mettle on the paradoxes of the blue sky. In 1849, he noted the problems of explaining it by light refracting through water droplets and proposed instead tiny bubbles suspended in air. Essentially, he was reviving Newton's theory of the "first-order blue," though now using the wave theory of light and bubbles in place of droplets.[37]

Nevertheless, in 1853 Ernst Wilhelm von Brücke argued that the blue from such bubbles, caused by interference as with Newton's rings, is not as saturated as the sky's blue. Here "saturation" has the technical meaning of the purity of the hue, the degree to which it appears to differ from a gray of equal brightness. Also, Clausius' bubble theory (like Newton's droplets) does not address the problem of polarization.[38]

Brücke's paper refers to Leonardo's *Treatise on Painting* and Goethe's color theory, showing the persistent fascination of Goethe's work especially in the German-speaking world that venerated his poetry. Brücke's main activity as a distinguished physician and physiologist shows also that amateurs were still making important contributions to physics in the middle of the nineteenth century. Indeed, Brücke helped to introduce physical and chemical methods into medical research. His investigations of how chameleons change color led to his observations about the sky's blue.

To make his arguments more vivid, Brücke dissolved mastic (a vegetable resin also called gum arabic) in alcohol to make insoluble droplets. He, too, tried his hand at bottling the sky. Brücke noted that his miniature droplet-filled atmosphere had a sky-blue color approaching the right purity (saturation). Saussure's bottle of ammonia dissolved in copper sulfate solution relies on a specific chemical reaction and depends on the light directly *transmitted* through the solution. Brücke's example uses light *scattered* indirectly and uses suspended droplets whose exact nature is not significant. His observation can be easily repeated with many different suspensions, including ordinary milk (see experiment 5.6 in appendix A). He also began to consider the significance of the size of the scattering particles compared to the wavelength of light, which later would prove to be critical.

Brücke objected to bubbles not only because they could not provide the right purity of blue or the observed polarization. Leaving aside the question of color saturation, we find that the sky is no one particular shade of blue. Here, the naked eye already teaches us much, if we are prepared to draw conclusions from sensitive observations. As Brücke noted, the eye registers the complex mixture of colors in the sky through the perceived degree of purity or saturation. In so doing, it perceives subtle qualities of light whose quantitative measurement came much later.

Ibn al-Haytham already knew that the sky's hue varies from horizon to zenith; Saussure and Humboldt measured its variation from day to day and place to place. Thus, the sky is not a monochromatic source of light but a synthesis of many different wavelengths. That is, a clear sky is blue *perceptually*, but not *spectroscopically*. By Brücke's time, physicists and astronomers had begun to measure the spectra of the sun and other stars.[39] Using prisms or diffraction gratings, they broke up the incoming light in order to observe the various wavelengths constituting it, which are characteristic of the elements making up the radiant source. By comparison with the sun, the ambient light of the sky is far dimmer, making it hard to get a clear, precise spectrum.

Astronomical observations of polarized light led to further insights. In 1860, the astronomer Gilberto Govi made a series of observations of the polarization of the light from the tails of comets. The particular polarization he observed had shown that those tails are gaseous, not solid. Govi decided to imitate this phenomenon in the laboratory. He directed a bright beam of sunlight through a small hole into a dark room filled with incense smoke. He observed that the scattered light was not only blue (as had been known at least since Leonardo's time) but definitely polarized. Moreover, this polarization was greatest at 90° from the incident beam, just as the light in the sky is most polarized 90° away from the sun. There were also neutral points of no polarization. Govi duplicated the experiment with tobacco smoke, with the same results, indicating that the phenomenon did not depend on the composition of the smoke. His experiments were analogous to Brücke's, though using airborne smoke particles, instead of droplets in liquid.[40]

Thus, scattering by small particles promised to explain the polarization of sky light. Here the concept of scattering allows a more general insight into the interaction of light with matter than did the older concepts of reflection or refraction. Govi noted that the 90° polarization he observed in light *scattered* through smoke would *not* occur if the light were *reflected* as if the smoke particles were tiny mirrors, which would cause a different angle of polarization upon reflection. He emphasized that a similar difficulty would apply to Bouguer's "molecular reflection" if molecules acted only as mirrors.[41] The problem, then, was to clarify how scattering by particles might differ from reflection by mirrors. Here lay the path to understanding the purity of sky blue.

6 True Blue

By the 1860s, the sky's blue had become something of a mystery. If it were caused by particles suspended in air, what could they be? Could they be water? Only 2 percent of the volume of the atmosphere, at most, is gaseous water vapor, compared to 1 percent for the inert gas argon. Such a tiny fraction could not reasonably account for the color of the whole. Of course, depending on the ambient humidity, small water droplets are present in the atmosphere but would evaporate on hot, dry days, when the sky is often a deep blue. Govi's experiments suggested some kind of particulate smoke. Even so, why should the artificial atmospheres created by Govi and Brücke look *blue*, rather than some other color?

Writing in 1862 to the physicist John Tyndall, the celebrated astronomer Sir John Herschel was perplexed about what could possibly explain the color and polarization of skylight:

> The cause of the polarization is evidently a reflection of the sun's light upon *something*. The question is, On what? Were the angle of maximum polarization 76°, we should look to water or ice as the reflecting body, however inconceivable the existence in a cloudless atmosphere on a hot summer's day of unevaporated molecules (particles?) of water. But though we were once of this opinion, careful observation has satisfied

> us that 90°, or thereabouts, is a correct angle, and that therefore, whatever be the body on which the light has been reflected, *if polarized by a single reflection*, the polarizing angle must be 45°, and the index of refraction, which is the tangent of that angle, unity; in other words, the reflection would require to be made *in* air *upon* air![1]

Thus, Herschel eliminates all the impossible cases, leaving only scattering on air itself. He considers this impossible "because there can be no reflexion upon air in contact with air of the same density." Both reflection and refraction depend on light experiencing a changing index of refraction, as when passing from air into water. But light going from air to air experiences no such change and thus cannot refract or reflect. Yet, as we have seen, water droplets are impossible too. Herschel also notes that "it is only where the purity of the blue sky is most absolute that the polarization is developed in its highest degree, and that where there is the slightest perceptible tendency to cirrus [clouds] it is materially impaired."

This is an elementary but devastating observation. If the slightest wisps of cloud *diminish* the polarization of the sky and also its perfect blueness, then the water in the clouds cannot be the cause of either, nor can any other particulate component of the clouds. Bearing Herschel's comments in mind, it is no wonder that Tyndall in 1869 wrote that "the blue colour of the sky, and the polarization of skylight . . . constitute, in the opinion of our most eminent authorities, the two great standing enigmas of meteorology."[2]

By that time, a number of experiments seemed to capture sky blue in earthly bottles. In 1845, well before Brücke's work, Sir John Herschel noted that light passing at certain angles through a transparent, colorless solution of quinine could give rise to a "beautiful celestial blue colour."[3] In 1866, Henry Roscoe showed that a dilute, slightly milky suspension of sulfur in water scattered blue light so as to give "an exact imitation of the condition of the atmosphere." He concluded that "the atmosphere is filled with particles which reflect the blue rays and transmit the red. What the exact nature of these particles may be, it is hard to say." Roscoe speculated that they may be "particles of soda" or perhaps "finely-divided extra-terrestrial meteoric dust."[4] But he did not explain how such particles manage to scatter light in this way, nor did he consider the issue of polarization.

Tyndall took important steps to address these problems. After finishing his doctoral work in physics and chemistry in Germany, he took a walking tour in Switzerland and later undertook several mountaineering expeditions. He was the first to climb the Weisshorn and one of the first to scale the Matterhorn. Like Leonardo before him, he climbed Monte Rosa and became deeply interested in the physics of glaciers and the color of the sky. Tyndall had been struck by Govi's findings and noticed the blue color of peat smoke in Ireland, of burning leaves in Hyde Park, and even of his own cigar. In all these examples, the blue light is polarized, like skylight. He also noted Hermann von Helmholtz's explanation that tiny suspended particles cause the blueness of the eyes.[5] It seems this beautiful human feature not only reminds one of the sky but is itself another miniature sky, also colored by scattering. In the end, this insight will be sustained, but only by showing that the eye's blue is scattering from something different than the cause of the sky's blue.

Tyndall began to investigate whether other aerosols—particles suspended in air—besides smoke might also show a blue light. Here, the wave theory of light is crucial. Already in 1852, the physicist George Gabriel Stokes showed that the "beautiful celestial blue color" of quinine solutions came from incident *violet* wavelengths passing through the solution and somehow shifting to blue light. He called this *fluorescence*, but similar phenomena had long been known, such as minerals that glow in the dark after exposure to light or yellow solutions that appear blue when held to light. Stokes argued that this was due to light exciting "vibratory movements among the ultimate molecules of sensitive substances." In contrast, he noted that light scattered from particles much smaller than its wavelength should be most strongly polarized *perpendicular* to its incident direction, in the plane defined by the light's transverse vibrations. He had verified this experimentally with hydrosols (particles suspended in water, like Brücke's). For the first time an explanation emerged for the transverse polarization of skylight as a direct result of the nature of light as a transverse wave. Though Stokes did not make these connections explicit, his work indicates new ways in which light can interact with matter, especially with molecules.[6]

Tyndall came to the study of extremely small particles independently, through his experimental work, though he gives credit to Stokes's earlier insight. Tyndall's studies of glaciers led him to use photochemical smogs to study the absorption of sunlight in the atmosphere. In 1868, he noticed that when he sent beams of light into his artificial smogs, a sky-blue light emerged. He wrote in his notebook a memo to "connect this blue with the colour of the sky." Not long after, he

> took a pleasure . . . in determining whether in all its bearings and phenomena the blue light was not identical with the light of the sky. This to the most minute detail appears to be the case. The incipient actinic clouds are to all intents and purposes pieces of artificial sky, and they furnish an experimental demonstration of the constitution of the real one.[7]

The quest to recreate the sky in a bottle was indicating answers to the enigma of the real sky. Where Govi used natural smoke, Tyndall used photochemistry "to render the chemical action of light upon vapours *visible.*" These photochemical reactions provided clouds of particles of various sizes.

To do this, Tyndall used a glass cylinder 90 cm long and 7 cm in diameter, which he filled with mixtures of various dilute gases and illuminated with a narrow beam from one of the new, powerful electric lamps (figure 6.1). He began with carbon dioxide gas, which he was trying to decompose with light. He saw "a faint bluish cloud" from which came light polarized at right angles to the incident direction of the beam. Since pure carbon dioxide would not have reacted photochemically in this way, Tyndall's samples were probably impure. But his basic result was not peculiar to that compound; Tyndall saw a similar blue cloud with many other substances: benzene, carbon disulphide (dry-cleaning fluid), butyl nitrite, among others.[8]

Since these volatile compounds are highly noxious, carcinogenic, and explosive, it is dangerous to experiment with them even in a specially equipped laboratory, and it is out of the question to do so at home. Experiment 6.1 in appendix A offers a safe alternative based on Roscoe's earlier approach. In comparison with Saussure's sky in a bottle, Tyndall's has an interesting element of time. Where Saussure's copper-ammonia precipitate immediately shows an unchanging, strong azure, Tyndall's takes a few minutes before a far more delicate blue cloud appears. At the same time, the light transmitted straight through the liquid becomes yellowish-orange, so that, looking first sideways and then along the beam, one sees both a miniature sky and a pale sunset. (This does not account for deep red sunsets, to which we will return in chapter 10.)

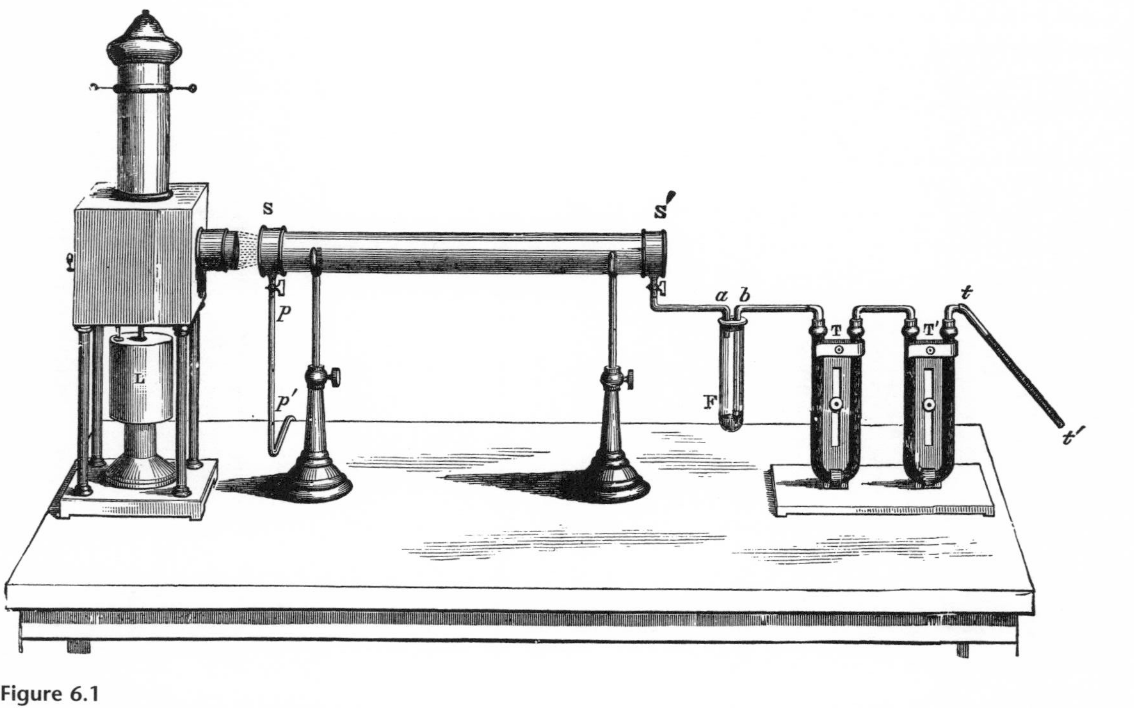

Figure 6.1

Tyndall's apparatus to demonstrate the "sky in a bottle" (1871). Light from an electric lamp (*L*) passes through a tube (*SS′*) filled with vapors from a tube (*F*) containing volatile liquid, over which air flows that has been filtered successively through cotton-wool (*tt′*), caustic potash (*T′*), and sulfuric acid (*T*) to remove floating matter and water vapor. The pipe *pp′* connects to a vacuum pump that can greatly reduce the pressure in the observation tube (*SS′*).

The delicate blue in Tyndall's bottle was quite transient, after a few minutes becoming a white cloud, from which the polarization disappeared. With different gases, Tyndall saw strange formations that looked like "roses, tulips, and sunflowers," moving and changing organically, including "clouds of a special pearly lustre" very like those he had seen in the Alps. He marveled at a color that "rivals, if it does not transcend, that of the deepest and purest Italian sky."[9]

The unfolding time scale of his experiments brings forward the crucial factor: the size of the particles. The blue cloud is at first formed of particles "whose diameters constitute but a very small fraction of the length of a wave of violet light."[10] Like Stokes, Tyndall understood the significance of this small size for the blue sky, for that blue disappeared when the cloud began to form into smaller particles that then coagulated into larger ones. Tyndall concluded that the blue cloud "is gradually rendered impure by the introduction of particles of too large a size—in other words, as real clouds begin to be formed." Tyndall's bottle held not just a blue sky but clouds emerging from it, as it seemed to him.[11]

These experiments struck John Ruskin, who had emphasized the deep significance of the sky as a touchstone of ultimate truth, both in art and in nature. He deepened this romantic theme in wide-ranging writings that shaped the sensibilities of many thoughtful people, including Marcel Proust, Leo Tolstoy, and Mahatma Gandhi. Ruskin's influential book *Modern Painters* included a chapter entitled "Of the Open Sky" (1843) that began by noting that "it is a strange thing how little in general people know about the sky. . . . there is not a moment of any day of our lives, when nature is not producing scene after scene, picture after picture, glory after glory," among which he dwells on the "simple open blue."[12] In his 1869 lectures, just two months after Tyndall publicly demonstrated his results, Ruskin noted that these experiments showed

> first, that the Greek conception of an aetherial element pervading space is justified by the closest reasoning of modern physicists; and, secondly, that the blue of the sky, hitherto thought to be caused by watery vapor, is, indeed, reflected from the divided air itself; so that the bright blue of the eyes of Athena, and the deep blue of her aegis, prove to be accurate mythic expressions of natural phenomena which it is an uttermost triumph of recent science to have revealed.[13]

Perhaps Ruskin's use of Tyndall's experiments smacks of special pleading to validate his mythic readings. Nevertheless, Ruskin claimed that the sky's blue comes "from the divided air itself," and not any particles suspended in it. He probably was unaware of Bouguer's ideas about molecular reflection, but he did believe in molecules and in 1884 described a transparent cloud appearing blue from "light reflected from its atoms."[14] As we will see, only in 1873 did Maxwell state privately the crucial relevance of the atomic theory to sky blue, which Lord Rayleigh finally confirmed twenty-six years later.

Ironically, though in this case Ruskin was suspicious of modern science, his imaginative conviction of the power of Athena led him to insist with uncanny prescience on what he called the "diffraction" of molecules in transparent air, a crucial process whose atomic basis Tyndall himself did not understand. In 1884, Ruskin teased Tyndall for asserting that air saturated with "transparent aqueous [water] vapor" caused the blue sky: How could what made air *transparent* also make it *blue,* were it not that air molecules themselves "diffract" the light? Here the visionary artist saw more clearly than the sober scientist.[15]

These questions had large implications for Ruskin. He considered that the true understanding and representation of the sky was fundamental to painting and to the human spirit. Accordingly, he invoked these new discoveries as essential evidence. Well before Tyndall's discoveries, Ruskin emphasized in *Modern Painters* that sky blue "is of course the color of the pure atmospheric air, not the aqueous [water] vapor, but the pure azote [nitrogen] and oxygen, and it is the total color of the whole mass of that air between us and the void of space."[16] From this, Ruskin draws important consequences for painting that he finds realized particularly in the paintings of J. M. W. Turner:

> And if you look intensely at the pure blue of a serene sky, you will see that there is a variety and fullness in its very repose. It is not flat dead color, but a deep, quivering, transparent body of penetrable air . . . and it is this trembling transparency which our great modern master [Turner] has especially aimed at and given . . . ; something which has no surface, and through which we can plunge far and farther, and without stay or end, into the profundity of space; whereas, with all the old landscape painters, except Claude [Lorraine], you may indeed go a long way before you come to the sky, but you will strike hard against it at last.[17]

Thus, Turner's artistic insight arrived at an understanding that Tyndall's experiments confirmed. In turn, Ruskin praised the scientist's artistry. "To form, 'within an experimental tube, a bit of more perfect sky than the sky itself!' here is magic of the finest sort." Science converges with art in "the true wonder of this piece of work." In response, Tyndall wrote that Ruskin's identification of Athena with air was "bold and true": "No higher value than this could be assigned to atmospheric oxygen."[18]

Yet this was a rare and passing moment of concord, for Ruskin thought Tyndall's science turned the living world, felt by human sensibility, into lifeless and thus untrue abstractions. This was not because Ruskin was ignorant of science. As a child, his greatest ambition was to become the president of the

Geological Society. Even before becoming an artist and critic, he had studied geology and was a passionate alpinist (though far less energetic and athletic than Tyndall). But compared with the new generation of theoretically inclined scientists, Ruskin's model was Saussure, who "had gone to the Alps, as I desired to go myself, only to *look* at them, and describe them as they were, loving them . . . more than himself, or than science, or than any theories of science." When Ruskin went to the Alps, he not only sketched the sky but measured its blue with his homemade cyanometer.[19]

Though often considered an arch-materialist, in 1868 Tyndall had put forward a much-noticed defense of the "scientific imagination" as the way in which science rises above pure materialism toward a more comprehensive theoretical vision. He particularly instances his work on sky blue. For him, the experimentalist pursues "the continued exercise of spiritual insight, and its incessant correction and realization. His experiments constitute a body, of which his purified intuitions are, as it were, the soul." This may have disturbed Ruskin, who considered the imagination to be the essential terrain that contemporary science violated by its reductive claims. On the other hand, Ruskin also admired science, at least as practiced by Saussure. The strength of Ruskin's convictions about science and the full tension between him and Tyndall emerged during a vehement (though one–sided) controversy they subsequently had about the movement of glaciers. In a marginal note in his copy of Tyndall, Ruskin called him "the entirely damned oaf and puppy."[20]

Then too, even Ruskin's 1869 praise of Tyndall's "sky in a bottle" was tempered by irony; he said enigmatically that he had "bitter reason to ask [the scientists] to teach us more than yet they have taught." The very day he wrote his praise of the experiment, Ruskin wrote to a friend that he had to "say how wonderful this putting this sky in a bottle is; and then say—for last word—that I'll thank *them*—the men of science—and so will a wiser future world—if they'll return to old magic—and let the sky out of the bottle again, and cork the devil *in*." Later, we will return to this request of Ruskin's.[21]

He went on to make a still darker prophecy: By capturing the sky in a bottle, science may leave a diabolical stench behind that will foul the air we breathe. He lectured in 1884 on an ominous "plague-cloud" that "dims the blue sky" and overshadows his century. Doubtless he was reflecting also the lurid skyscapes caused by the explosion of Krakatau (Krakatoa) in 1883, wondered at worldwide. But Ruskin drew an ominous prophecy from the "poisonous smoke" he saw belching from "at least two hundred furnace chimneys in a square of two miles on every side of me." This was no mere smoke but "looks more to me as if it were made of dead men's souls."[22] We cannot forget Ruskin's indictment now, when such pollution shrouds the world. To what extent does science deserve blame if greedy industry and heedless consumerism have abused its powers? Yet beyond such questions of blame, it is hard to imagine any solution that does not employ those same powers.

Even in his bitter irony, Ruskin was aware that he had to address and engage science, not simply denounce it. Consider his correspondence with the physicist Oliver Lodge, who in 1885 argued that sky blue was caused by light scattering from "fine impalpable dust," which could serve as nuclei for condensing water droplets. Lodge tried to explain the kinetic theory of incessantly moving molecules, which made Ruskin "sick and giddy." But ironically, the atomist Lodge could not see the possibility of purely molecular scattering and fell back on the old dust theory, which Ruskin thought he assumed without any real demonstration: "I don't believe in them yet!—except in Tyndall's experiments at the Royal Institution." Because Tyndall produced blue light from gas filtered through cotton to remove dust, Ruskin concluded that air by itself was sufficient. He gave no account of how the air could act in this way, beyond pointing to the parallel with ancient Greek myths; he found no enlightenment in the "mystic motions" of molecules.[23]

For his part, in his 1869 paper Tyndall certainly did not advocate this radical view. Ironically, he was a convinced atomist who palpably "saw the atoms and molecules, and felt their pushes and pulls," as Thomas Huxley put it.[24] Nevertheless, from his experiments Tyndall concluded that air was "optically empty," as he put it, lacking any ability to scatter light by itself, as far as he could tell, after he had removed microscopic dust particles in his cotton filter. He therefore continued to believe in some sort of extremely small particles, whether water droplets or some other kind of submicroscopic "sky matter," perhaps even "floating organic germs." As an important advocate of the germ theory of disease, Tyndall was aware of the omnipresence of airborne microbes.[25] A light beam stabbing through a darkened room reveals "the floating dust of the air, which, thus illuminated and observed, is as palpable to sense as dust or powder placed on the palm of the hand." But he also knew that such dust gradually settles "until finally, by this self-cleaning process, the air is entirely freed from mechanically suspended matter."[26] He did not acknowledge the difficulty this poses for any stable sky matter.

As the letters in appendix B show, Tyndall confronted another alternative: the blue of the sky may come from "a minute reflexion even from the ultimate molecules," as Stokes put it. This suggestion ultimately goes back to Bouguer in 1758. Tyndall wondered about his artificial skies: "*what must be the size of the particles when they first appear?*" Though he is on the verge of answering that they probably begin not much bigger than molecules, he draws back: "I suppose we must conclude that even then each particle is a heap of molecules and each molecule a heap of atoms."[27] Both Stokes and he seemed convinced by the evidence that small particles in the artificial sky were enough to give a convincing blue. Further, Tyndall took the fact that molecules are "incapable of being . . . seen themselves" to prove that they are therefore incapable "of scattering any sensible portion of light which impinges on them." Accordingly, he hoped that "we may in time find out how many of the molecules thus grouped together are competent, under the conditions referred to, to send a sensible amount of light to the eye."[28] Though Tyndall felt strongly that molecular reflection "will prove to be an hypothesis that has not a single fact to support it," Stokes noted presciently that disproving it was "hardly within the reach of experiment considering how small are the thicknesses of air looked through in our experiments compared with the miles through which we look in regarding the sky."[29]

Tyndall emphasized the universality of what he had found: The blue cloud emerged from many different chemical reactions and pointed to something more fundamental than any one of them. The issue of polarization sharpens this. Brewster had long ago pointed out that the angle of complete polarization of reflected light depends crucially on the substance, different for water (54°) or diamond (67°). Yet the blue cloud, and the blue sky, have a universal polarization of 90°, independent of their material. Tyndall concluded that "the law of Brewster does not apply to matter in this condition; and it rests with the undulatory [wave] theory to explain why."[30]

The traditional ray theory of optics considered light as made up of "beams" that refract or reflect geometrically. In the wave theory, light is not geometric beams but extended wave fronts, whose size is given by their wavelength. When light impinges on a piece of matter that is larger than its wavelength, we must take into account the collective effect of scattering from the enormous number of constituent particles that make up the matter, each responding differently to the wave. Under certain conditions, this can result in what we call specular (mirror-like) reflection or refraction. If light travels through an optical medium that is sufficiently homogeneous and arrives at a large, smooth interface with another medium, reflection and refraction can occur. Here, the scale of size, smoothness, and homogeneity is given by the wavelength. For instance, this describes light reflecting and refracting as it impinges on a water surface. Where these conditions of size, smoothness, and homogeneity do not apply, more complex forms of scattering will result. We do not see reflections in clouds or dirt, which do not provide a smooth interface, on the scale of wavelength. In contrast, a mirror is ground and silvered exactly in order to have sufficient smoothness to reflect a clear image.[31]

Thus, scattering includes reflection and refraction, under the conditions just given. But scattering also includes phenomena that did not fall under these special conditions. When the particles are much smaller than the incident wavelength and there are no such smooth surfaces and interfaces, we have to find a new way to describe the scattering.

Tyndall uses metaphor to try to distinguish these kinds of scattering. First he describes how a cloud of water droplets scatters sunlight, which he considers to be "waves of ether." In one limit, the scattering particles are much larger than the wavelength of light. (Note the way he seems to use "reflection" and "scattering" interchangeably.)

The cloud, in fact, takes no note of size of the part of the waves of ether, but reflects them all alike. Now the cause of this may be that the cloud particles are so large in comparison with the size of the waves of ether as to scatter them all indifferently. A broad cliff reflects an Atlantic roller as easily as a ripple produced by a sea-bird's wing; and in the presence of large reflecting surfaces, the existing difference of magnitude among the waves of ether may disappear.[32]

That is, Tyndall argues that the cloud is white because it scatters all wavelengths equally. Now consider the other limit:

But supposing the reflecting particles, instead of being very large, to be very small in comparison with the size of the waves. Then, instead of the whole wave being fronted and in great part thrown back, a small portion only is shivered off by the obstacle. Suppose, then, such minute foreign particles to be diffusing in our atmosphere. Waves of all sizes impinge upon them, and at every collision a portion of the impinging wave is struck off. All the waves of the spectrum, from the extreme red to the extreme violet, are thus acted upon; but in what proportions will they be scattered? Largeness is a thing of relation; and the smaller the wave, the greater is the relative size of any particle on which the wave impinges, and the greater also the relative reflection.

For the first time, a physical argument is given for why shorter wavelength (bluer) light would be scattered more than longer (redder). Small particles represent a larger obstacle to the shorter wavelengths than to the longer. Tyndall amplifies this aquatic analogy, tacitly using "largeness" to describe the wavelength, not the wave's amplitude (height):

A small pebble placed in the way of the ring-ripples produced by heavy rain-drops on a tranquil pond will throw back a large fraction of each ripple incident upon it, while the fractional part of a larger wave thrown back by the same pebble might be infinitesimal. Now to preserve the solar light white, its constituent proportions must not be altered; but in the scattering of the light by these very small particles we see that the proportions *are* altered. The smaller waves are in excess, and, as a consequence, in the scattered light blue will be the predominant colour. The other colours of the spectrum must, to some extent, be associated with the blue: they are not absent, but deficient. We ought, in fact, to have them all, but in diminishing proportions, from the violet to the red.

Recall that Brücke's observation of the saturated quality of the sky's color led us to consider the possibility that it has no single wavelength, but is a complex mixture. Here Tyndall has given an explanation that requires this. But his argument also requires that violet wavelengths, being shorter than blue, should be scattered more. Why, then, is the sky not *violet*? Though one of his correspondents asked him about it, Tyndall did not seem to notice this violet puzzle, to which we will return later.[33]

Tyndall's reliance on analogy and physical intuition recalls the work of Michael Faraday, the experimental genius who had preceded him in his post at the Royal Institution in London and about whom Tyndall wrote a sensitive biography.[34] Both were inspired experimenters, not mathematically inclined, with a gift for visualizing the physical crux of the matter. Both began in humble backgrounds and only by dint of extraordinary labor became eminent public figures in science. To be sure, Faraday's contributions were greater in his seminal experiments and theoretical insights into electromagnetism. Yet Tyndall's experiments and deductions about the blue sky have a special beauty all their own.

Indeed, here he excelled his master, for in 1854 Faraday confessed that, though "my own knowledge does not render me competent to give an opinion" on the cause of the blue sky, he deferred to "a high mathematician," Clausius, and accepted the bubble theory. It is ironic that Faraday also argued convincingly that there was insufficient water in the atmosphere to cause the blue color; he calculated that if all the atmospheric water were condensed into the liquid state, it would not make a layer of more than a few inches in depth, "and I leave you to judge how utterly insufficient this would be to produce the blue skies seen in this country, much less those of Italy and other parts of the world. Three inches in depth of the Rhine water at Geneva, which is as blue as any water I know of, if held in a glass vessel between the eye & the sky would give scarcely an appreciable effect of colour." This sharp observation is characteristic of him, though not his attempt to accept the bubbles "*not because of any colour they have in themselves*, but because of their optical effect on the rays of light passing through the atmosphere." Here Faraday seems to have ignored his own argument about the available quantity of atmospheric water. He also seems unaware of Brücke's objection and surprisingly deferential towards Clausius' mathematical authority. Faraday's deepest reason may have been that he thought "he had actually seen the cloud-vesicles," as he confided to Tyndall. In turn, Tyndall long "sought for these acqueous bladders" in the Alps, but never was able to find them.[35]

Faraday himself could not read a single equation and made his greatest discoveries by following out his own highly physical, visual reasoning. But what he achieved so brilliantly elsewhere eluded him in the problem of the blue sky. His electromagnetic theory was given precise mathematical form by James Clerk Maxwell, who admired and followed Faraday's vision. Tyndall's work inspired the theoretical synthesis of John

William Strutt, the third Baron Rayleigh, who deeply reinterpreted Tyndall's original insight. Rayleigh is not as well known as the famous innovators like Maxwell, in his own time, or Einstein and Planck in the generation after him. He did not make any revolutionary breakthroughs, as these others did. Yet his superb mathematical craft and physical insight made him a physicist's physicist. More than a century after it was written, his *Theory of Sound* remains a masterpiece on the physics of waves.

Rayleigh had studied Tyndall's writings and had also been interested in the color of the sky while studying Maxwell's ideas about color vision. As a young man, Maxwell had developed a color top, in which patches of red, blue, and green paper could be spun so as to create the perception of other colors (figure 6.2). Maxwell revived Young's three-receptor theory of color vision and created the modern science of quantitative colorimetry. He also exhibited the first color photograph (1861), made by superimposing three "color-separated" images, each a black-and-white photograph made through a red, blue, or green filter and re-projected through the corresponding filter. This dramatic demonstration gave further evidence that the brain can construct a full color image from only three wavelengths. Studying Maxwell's work, Rayleigh noticed that the matching of colors is affected by the color of the sky at the time of observation, "and this led him to make comparisons of the colour of the blue sky with that of direct sunlight," according to his son.[36]

In 1871, Rayleigh first addressed Tyndall's work. In general, he agrees with Tyndall's approach, aiming to make it more quantitative and theoretically grounded. He begins by answering Tyndall's question about why Brewster's law seemed to break down for the tiny particles. This Rayleigh explains easily;

Figure 6.2

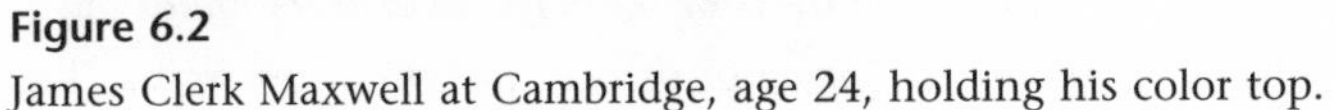

James Clerk Maxwell at Cambridge, age 24, holding his color top.

Brewster's law concerns reflection from smooth surfaces, but if the particles are very small, their curvature is much smaller than the wavelength so that the laws of specular reflection do not apply. As Rayleigh thinks about the basic process of light scattering, following the general preconceptions of his time, he assumes that there is an ether in which the small particles are immersed, so that they "*load* the aether so as to increase its *inertia*."[37] That is, the particles act as tiny weights that effectively increase the inertia of the ether. In effect, the picture is rather like an old-fashioned set of bed springs, with little masses attached to the springs. Then Rayleigh applies Newton's third law: to every action, there is an equal and opposite reaction. Thus, the force that a particle exerts on the ether is equal and opposite to the force the ether exerts on the particle.

Rayleigh's argument is very general, assuming only Newton's laws and a simple picture of the particles and ether moving in response to a simple incident wave. He invokes no electrical property of the scatterers, only that they respond to incoming waves as would an elastic solid. Rayleigh gives a Newtonian argument for his main result but presents it even more simply using "dimensional analysis." This is an ingenious device introduced by Maxwell, in which consideration of the dimensions (such as length, mass, time) of an unknown quantity can give its proportionality to the factors on which it depends.[38]

Using this device, Rayleigh derived in a few simple steps the basic law governing the scattering of light from small particles. The details are given in the notes.[39] The crux of Rayleigh's scattering law is that *the brightness of scattered light from small particles falls off as the fourth power of the wavelength of the light.* That is, the longer the wavelength of the incident light (redder), the less it will be scattered. Likewise, the smaller the wavelength (bluer), the more it will be scattered. For instance, red light of wavelength 650 nm (one nanometer = one billionth of a meter) and violet light 400 nm have a ratio of wavelengths 650/400 = 1.63. Then, because of the fourth power, we expect the brightness of scattered violet light to be seven times that of red (since $(1.63)^4 = 6.97$). Similar ratios apply to all other wavelengths in between. This, at last, gives a quantitative account of Tyndall's qualitative argument, though it still seems to indicate that the sky should be violet.

Rayleigh carried his argument further to address the alternative of bubbles or thin films, rather than particles. He again used dimensional analysis to show that the brightness of the light scattered from films or bubbles should go inversely as the *square* of the wavelength, rather than inversely as the fourth power for scattering from particles.[40] To return to our example, the brightness of a violet wavelength scattering from bubbles is only a factor of $(1.63)^2 = 2.64$ times that of a red wavelength, compared to a corresponding factor of $(1.63)^4 = 6.97$ for scattering from particles. Thus, if the sky's color were due to bubbles, we would expect its violet to be about a factor of three (6.97/2.64 = 2.64) times less bright than if due to particle scattering.

These predictions now invite quantitative tests of these theories. Indeed, Rayleigh was the first to measure the spectrum of the blue sky, in order to decide the issue.[41] His approach was simple: he compared incident sunlight to scattered skylight. On a sunny day with blue skies, he allowed the sun's light into a dark chamber through a narrow slit. Then that light passed through prisms that would break it up into its constituent wavelengths, one of which could be selected by a screen with a slit that would block out the others. This would illuminate a sheet of paper, so that he could observe it without damaging his eyes, and would give Rayleigh a standard for the brightness of the incident sunlight of that color.

Next to this, Rayleigh set up a similar apparatus to bring the light from a patch of blue sky through another slit, then through prisms and slits that would isolate the same wavelength of the scattered light. Finally, he would compare by eye the brightness of the incident and scattered light, using lenses and eyepieces so that he could more easily have both lights before him. Because Rayleigh was using his eye to judge and compare, his measurements inherently included the eye's variable sensitivity to light of different wavelengths. Even so, he was able to get very reliable results (figure 6.3). Among six sets of

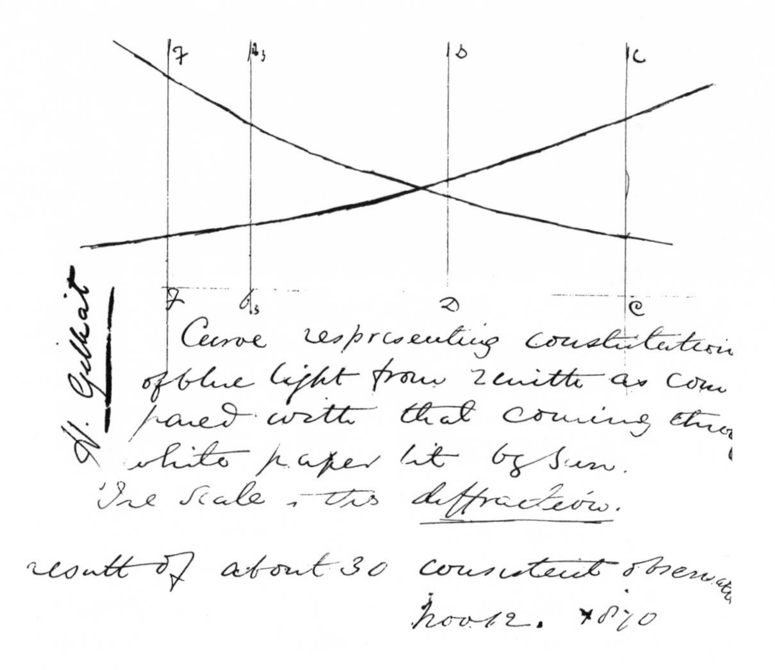

Figure 6.3

Lord Rayleigh's sketch of his first observations on the blue of the sky. His handwritten text reads: "Curve representing constitution of blue light from zenith as compared with that coming thru white paper lit by sun. The scale is the *diffraction*. Result of about 30 consistent observations. Nov. 12 1870." The letters give the wavelengths of light by referring to spectral lines of different elements, such as *F* (blue hydrogen line at 486.1 nm), *D* (yellow sodium line at 589.3 nm), and *C* (red hydrogen line at 656.3 nm). The curve beginning at the upper left represents the zenith sky; the other shows sunlight diffused through paper. Rayleigh made this sketch on the back of a mathematics examination paper by "H. Gilliat." (Courtesy Rayleigh Archives.)

observations, his results varied only about 2 percent. With intermediate colors, his results tallied very closely (within a few percent) with his law of scattering from small particles. Between the smallest and the largest wavelengths, he predicted a ratio of 3.2 and observed 3.6. Because his measurements did not extend past blue wavelengths into the violet, he did not address the violet puzzle.

Nevertheless, the sky was even bluer than he expected. To be sure, the measurements in the less luminous parts of the spectrum were more prone to error, since they involved the faintest observations, the hardest to compare accurately. Rayleigh also noted that his standard of light was not simply sunlight but included a certain amount of "earthshine," light scattered back into the sky from the earth itself (recall the speculations of Lambert and Noellet about such reflected light as the cause of the sky's blue, here both answered and included).

These results finally showed quantitatively and observationally that the bubble theory definitely did not give a blue enough sky. There remains the unasked question: Why not violet? But a still more urgent puzzle also remained. At the end of his paper, Rayleigh notes the impossibility of the particles being water, given the heat of a summer's day, for "it is difficult to imagine particles of water smaller than the wave-length [of light] endowed with any stability."[42] The problem here is not so much that they are falling under gravity, for the terminal fall velocities of such small particles is very small, much less than the velocities of the updrafts they experience. But such droplets are prone to evaporation and also to coagulation with other droplets, resulting in larger droplets.[43]

Rayleigh ends with some diffident speculations; perhaps the particles at high altitude are different from those near the surface of the earth: "If it were at all probable that the particles are all of one kind, it seems to me that a strong case might be made out for common salt. Be that as it may, the optical phenomena can give us no clue." Rayleigh proposes salt because of its commonness, but in fact he knows he is only speculating. Even at the moment his new law of scattering triumphed, the mystery of the particles remained deeper than ever.

7 Blue Laws

Rayleigh went on to enlarge and refine his calculations, but his basic conclusions remained intact, giving no further insight into the nature of the particles themselves. At the time, his papers seemed abstruse (as his son later recalled), but they pleased Maxwell, who wrote to a friend that "I think that Strutt [Rayleigh] on sky blue is very good. It settles Clausius' vesicular [bubble] theory." Tyndall also wrote to Rayleigh, smarting a little from his criticisms, but encouraging him to keep going.[1]

Ten years later (1881), Rayleigh returned to consider the interaction of light and particles according to Maxwell's mathematical theory of electric and magnetic fields. His attention had been drawn to this by Maxwell himself, who wrote him a letter in 1873 suggesting that Rayleigh calculate the scattering by particles small enough to be molecules. Maxwell added that "the object of the enquiry is, of course, to obtain data about the size of the molecules of air."[2]

Here we recall Bouguer's suggestion that sunlight might "reflect" from the molecules of the atmosphere to produce sky blue, though he did not realize that he was extrapolating the behavior of mirrors to particles smaller than the wavelength of light. Tyndall, Stokes, and Rayleigh thought that small particles might scatter the light instead. Maxwell finally identified the scatterers as molecules and also realized that their pattern of scattering would gauge their size.

To gain perspective on this insight, recall how peculiar the idea of scattering "from air to air," without change of medium, seemed to Herschel even in the 1860s. Tyndall remedied this by immersing the "other medium" in the air, spread out in tiny packets as the scattering particles themselves. Also, Rayleigh's arguments did not depend on the composition of the particles or even on their exact size, so long as they were much smaller than the wavelength of light.

Recall Helmholtz's suggestion that scattering from a dilute suspension of eye pigment gives blue eyes their color, whereas a dense suspension leads to brown. But the sky's color cannot be explained in this way. Water droplets evaporate; solids like salt or dust fall out of suspension. To address this problem, Lodge thought that incoming meteorites could replenish the dust that fell from the air. But the incoming mass is far from sufficient, as he could have calculated had he considered contemporary information about meteorites.[3]

Thus, Maxwell's view that these mysterious particles were simply the molecules of air itself seems obvious in retrospect. Yet it took some boldness to propose this. Here the atomic theory enters the story at a fundamental level. John Dalton outlined the modern atomic theory in 1808, and, by the late nineteenth century, most chemists accepted it. But there remained enduring controversy about the physical reality of atoms, apart from their utility as an explanatory device. The great chemist Humphrey Davy thought matter was composed of tiny particles but doubted they could be identified with the chemical elements. In 1833, John Herschel thought of Dalton's atoms as inspired guesses that a deeper theory might overthrow. Writing in 1870, Tyndall noted that "many chemists of the present day refuse to speak of atoms and molecules as real things."[4]

In the early twentieth century, the eminent chemist Wilhelm Ostwald thought energy was more basic than atoms. Ernst Mach died in 1916 still convinced that atoms did not exist as physical structures, even though he had seen with his own eyes radioactive scintillations that strongly indicate an atomic substratum.

These doubts were less unreasonable then than they might seem now, when most people do not question the existence of atoms, though they would be hard pressed to justify such a belief. After all, in the eighteenth century, some natural philosophers considered heat to be a physical substance called "caloric," because heat flows somewhat like a fluid. But after Lavoisier showed that caloric would have to have no weight or even negative weight in some situations, the consensus shifted: Heat is not a substance, but a state of motion whose measure is temperature. On the analogy of caloric, might atoms be not real substances, but states of organization or motion?

Maxwell took a leading role in convincing others in the physical reality of atoms. He helped shape thermodynamics, the study of heat and work, into the story of atoms in motion, which he called kinetic theory and which later was developed into statistical mechanics. He showed how the average velocity of atoms in a gas relates to the measured pressure and temperature. Even though one could not see or feel a single atom, myriads of them could hit with palpable force. Maxwell was hard at work on this in the years before Rayleigh's 1871 paper; in that year, Maxwell's *Theory of Heat* appeared and offered a clear exposition of kinetic theory intended for intelligent general readers.[5] So it was natural for Maxwell to suggest that Rayleigh investigate how light would pass through an assemblage of atoms: "If you can give us the quantity of light scattered in a given direction by a stratum of a certain density and thickness, . . . we might get a little more information about these little bodies."[6]

In his letter to Rayleigh, Maxwell goes on to note that "I make the diameter of molecules about 1/1000 of a wavelength," using his kinetic studies of molecules, as will be discussed shortly. Here Maxwell shows how seriously he takes their physical reality by giving them an actual size. Yet he was aware of how uncertain such estimates still were and was seeking to refine them. To be sure, Maxwell still believed in the omnipresence of an extremely tenuous ether, whose vibrations are the electromagnetic waves we call light, as well as infrared and ultraviolet radiation. He took Tyndall's and Rayleigh's particles and shrank them down to atomic size, still "loading" the ether with their inertia, as Rayleigh had.

In the process, Maxwell was developing a new rationale for atoms, though he did not make it explicit. If the blueness of the sky depends on scattering from small particles and no other substance is available for the suspension, then the solution to the mystery is that *those particles must be air molecules themselves.* Looking back, this deduction is startling in its simplicity and depth: *The physical reality of atoms is necessary for the blue sky.* This is a surprising and far-reaching conclusion to reach merely from noting the sky's color. Thus, both Maxwell and we would like further substantiation of this bold surmise. (We will also reconsider it in chapter 10.)

Rayleigh took many years to respond to Maxwell's 1873 challenge. His full answer appeared only in 1899 and affirmed Maxwell's insight with surprisingly little change to Rayleigh's original calculation, which assumed the existence of small particles, though not necessarily atom-sized.[7] In 1890, Ludvig Lorenz published essentially the same calculation, but his paper remained unnoticed at the time because it was in Danish.[8] Also, Lorenz did not connect his calculation with the problem of sky blue.

As in his earlier treatment of scattering, in his response to Maxwell Rayleigh does not need to consider the electromagnetic properties of atoms but only their behavior as an "elastic solid," as he puts it. His calculation proceeds as before, except that he needs to introduce the number of molecules per unit volume of air, in order to gauge the number of independent particles scattering the sunlight. This quantity will assume a central importance.

This number is closely related to one first considered by Amedeo Avogadro, a natural philosopher and teacher who was trying to make the precepts of the atomic theory more vivid. By 1808, Joseph-Louis Gay-Lussac had shown that, when different gases combine chemically, their volumes are in simple whole-number ratios. In 1811, Avogadro interpreted this to mean that equal volumes of any gas ought to contain an equal number of atoms (or of molecules), always assuming standard conditions of temperature and pressure (0°C at one atmosphere). For instance, if the volumes of two gases combine in a 2:1 ratio, the only way this can be reconciled with molecular theory is that equal volumes of each gas have the same number of molecules, so that two molecules of the first gas combine with one of the second, as two molecules of hydrogen (H_2) and one of oxygen (O_2) combine to form two molecules of H_2O. Thus, a cubic meter of hydrogen gas has as many molecules as a cubic meter of oxygen molecules. Because of their different molecular weights, the masses in each of these equal volumes (each "mole") are in the ratio of 2:32. Avogadro felt that emphasizing the equality in number of molecules in each mole would help students think about chemical reactions molecule by molecule, rather than gram by gram.[9]

But Avogadro could not determine the number of molecules per one mole because at that time chemistry had no way of counting the individual molecules. Not knowing that number meant that he did not know the size of one molecule, if the number of molecules in one cubic meter of a liquid or solid equals that volume divided by the volume occupied by a single molecule. Conversely, determining the size of atoms and molecules would fix the number.

The problem was difficult; in 1815, Young tried to solve it by considering the surface tension of liquids. But that meant he had to contend with extricating molecular sizes from the subtleties of surface tension. As a result, his calculations yielded a diameter for water molecules that is about ten times too small, compared to other, later determinations. This by itself was a good first step; the problem was how to check and correct this number by finding a method not dependent on dealing with something as complex as surface tension.

Not until 1865 did Josef Loschmidt give the first such measurement of the dimensions of the atom, which can be converted to "Loschmidt's number" of atoms per cubic meter, as it is still known in the German-speaking world. His method involved the use of Maxwell's kinetic theory to relate the densities of liquefied gases and the mean free path of molecules in the gas itself (meaning the average distance one molecule goes before colliding with another, roughly 66 nm in air at room temperature and sea level). Loschmidt deduced from this an estimate of the diameter of an air molecule, about 1 nm, three times the currently accepted value; he did not explicitly calculate the number per unit volume. Soon afterward, Johnstone Stoney

(1868) and William Thomson (1870) independently made similar calculations.[10]

Maxwell used the interdiffusion of different gases to give more precise and specific data. He was able to give a more accurate treatment of the kinetics of the molecules than Loschmidt had gotten from Clausius. Maxwell also made more accurate experimental measurements of the diffusion constants, working in collaboration with his wife. By 1873, he had arrived at molecular diameters ranging from 0.58 nm (for hydrogen) to 0.93 nm (carbon dioxide). The modern term "Avogadro's number" (N_A) refers to the number of molecules in one mole (to use the modern convention, this means exactly 12 grams of carbon-12 or the gram-equivalent weight for any other molecule, in terms of carbon-12). Maxwell's results corresponds to the number 4.1×10^{23} molecules per mole, compared to the modern value of 6.02×10^{23} molecules per mole. This 1873 determination proved more accurate than the value that Ludwig Boltzmann gave as late as 1893.[11]

This crucial number enters into Rayleigh's calculations of the scattering of light in the sky. At first, both he and Maxwell took Avogadro's number from these studies of gases, their diffusion properties, and liquid densities. Rayleigh also used Bouguer's measurements to show that the same value of N_A would account for both the brightness of the sky and the dimming of a star's light, within the limits of observational error. This was an important step, for it showed that the same theory of light scattering could account for several different phenomena.

Here Rayleigh went one step further. His fully developed formula related the extinction (scattering) of light to its wavelength, the index of refraction of the atmosphere, and Avogadro's number. Why not reverse this argument, taking data about atmospheric scattering and working backward to find

Avogadro's number? Using the data at his disposal (Bouguer's measurements), he found a value for N_A that agreed pretty well with Maxwell's values. In 1904, Thomson (by then Lord Kelvin) refined these values using 1901 observations made on Monte Rosa and Mount Etna of the ratio of the blue sky's brightness to that of sunlight. Kelvin's final result for N_A was about 2×10^{24}, a factor three times larger than the accepted value today.[12]

Though these precise measurements need exacting techniques, it is possible to estimate Avogadro's number using naked-eye measurements of the blue sky without elaborate equipment. To do so, we can measure the *visual range*, the minimum distance at which there is insufficient contrast to distinguish a black object from the horizon sky. At this distance, the scattered daylight in the column of air between observer and object is as radiant as the sky; past that point, the object is indistinguishable from the horizon. The visual range is an important measure of the clarity of the air, hence also of pollution; a number of modern instruments can measure it electronically to a high degree of accuracy. Yet naked-eye techniques can give a rough determination.

The simplest estimate of the visual range is how far one can see in the clearest weather. In 1760, Bouguer had estimated this at forty-five leagues (180 km).[13] In his 1899 paper, Rayleigh recalled seeing Mount Everest "fairly bright" from Darjeeling, one hundred miles (160 km) away. That same year, Kelvin watched for the "earliest instantaneous light of sunrise through very clear air" over Mont Blanc (68 km distant) and noted a blue flash that quickly "became dazzlingly white."[14] It is still possible to see great distances through unpolluted air. From my home in Santa Fe, I can just discern Mount Taylor (160 km) in the distant haze.

Now we can use these estimates of the visual range in Rayleigh's formula, which involves the index of refraction of air and the range of wavelengths of visible light. As the notes explain in more detail, a visual range of 160 km gives values for Avogadro's number ranging between 1–8 × 10^{23}, the right order of magnitude compared to the accepted value (6 × 10^{23}). It is striking that we can get this close from a crude measurement.[15] This requires sighting large, distant objects, but there are alternatives that involve only local observations. You can also use a cardboard tube and a piece of blackened glass to find Avogadro's number by looking at objects a few hundred meters away. The notes show how this leads to values for N_A of about 11 × 10^{23}, again only a rough estimate because many factors of visual judgment and air conditions enter in.[16] The visual range would surely be higher were not dust and other particulate matter in the sky, for these cause a considerable amount of scattering beyond that due to the air molecules themselves. The polarization of the sky also depends on atmospheric conditions. We will return to these differences between the actual sky and Rayleigh's idealizations in chapter 10.

Rayleigh's law of scattering was a critical step in the study of the brightness and color of the sky, giving a quantitative connection between observation and molecular reality. We have just used it to deduce the size of moleules from the brightness of scattered skylight. Rayleigh's successors took this further, among them his son, Robert John Strutt, later the fourth Baron Rayleigh.[17] The younger Rayleigh undertook direct measurements of the transparency of air in the laboratory, aiming to determine accurately the quantities we have estimated crudely above. This proved very difficult; recall that Tyndall concluded that pure air is "optically empty," leading him to postulate suspended particles to scatter the sunlight. The observed atmospheric scattering happens only when huge depths of air are involved, on the order of many kilometers, as we have seen. That meant the effects sought in the laboratory were of the magnitude of a few parts per million.

The elder Rayleigh "was not disposed to think it feasible" to measure these effects directly, recalled his son, who never quite succeeded himself. Indeed, most of the initial attempts at the beginning of the twentieth century failed, including those of Robert Wood (1902). Jean Cabannes finally succeeded in 1915, noting proudly that "the experiments confirmed the theoretical views and I was thus able to reproduce for the first time the blue of the sky in the laboratory." His greatest difficulty had been "to eliminate all extraneous light from the field of observation."[18]

Despite his disappointment, the younger Rayleigh made an important discovery about asymmetries in the polarization of scattered light, which has a more complex pattern because molecules are not symmetric spheres. His father was greatly interested, for he knew that the polarization of sky light, when examined carefully, also shows such asymmetries. This moved the elder Rayleigh to make new calculations in 1918 of the interaction of light as an electromagnetic wave with particles of more general shape. These calculations allowed more detailed comparison of experiment with theory. In the main, his basic theory stood confirmed, though these more exacting measurements also revealed both the irregular shapes of air molecules and the persistence of dust particles even in the cleanest, driest air.

During this same period, these measurements of atmospheric scattering took their place in the final battle to establish the atomic theory.

8 Blue Riders

After Rayleigh, the light from the sky could not be understood apart from atomic theory. Connecting the two was part of a many-sided effort to understand how atoms manifest their physical reality. As such, it drew the attention of the young Albert Einstein, who became interested in finding all possible ways of measuring Avogadro's number and hence molecular dimensions from diverse physical and chemical experiments. If this number were always the same, no matter what way it was measured, that would give critical confirmation to the atomic hypothesis. Indeed, by 1909 a dozen different and independent ways of measuring Avogadro's number had all converged on a value between 6 and 9×10^{23}.[1] In the previous chapter, we discussed the way skylight leads to one of these values. But if skylight really depends so deeply on atomic theory, we need to consider further what that theory implies. By considering some of the other ways of finding Avogadro's number, we will be able to return to skylight with a more complete perspective. A strangely varied chain of phenomena will lead us back to sky blue.

Let us begin with Einstein's doctoral work (1905), in which he determined Avogadro's number from the osmosis of sugar solutions. The resulting paper (1906) still remains one of his most cited works, for it proved to be of great practical use in the dairy and construction industries and also in ecology. This may surprise those who think of Einstein as a refined dreamer.[2] The common link between these oddly diverse activities is the behavior of solutions having suspended particles in various concentrations.

In the 1880s, Jacobus van 't Hoff had discovered osmotic pressure, the force exerted on a semipermeable membrane between two liquid solutions of different concentrations. He formulated an analogy between this pressure and the pressure exerted by a gas, showing that liquids, no less than gases, could be understood by atomic theory. Einstein took up this analogy and, through an ingenious argument, showed how measuring the diffusion and viscosity coefficients of a solution could allow determination of Avogadro's number. From the available data, in 1911 he was able to give a value of $N_A = 6.6 \times 10^{23}$.

Only eleven days after he completed his doctoral thesis, Einstein submitted a paper on Brownian motion that led to another way of determining N_A. In 1827, Robert Brown had studied the random jiggling of tiny particles he found contained in pollen, observed under a microscope. Gradually, he realized that despite their ceaseless dancing these particles were not living things, and that any particle would dance similarly, if only it were sufficiently small. During the nineteenth century, it became clearer that these motions were the result of the particle being hit by surrounding atoms about 10^{20} times a second, which often would push it randomly more to one side than the other. Einstein now applied his ideas about solutions to these dancing particles. Again using arguments based on diffusion, he derived a relation connecting the average jiggling of the particle with the viscosity of the surrounding medium and

Avogadro's number. He did not obtain a numerical value for this, but hoped that "some researcher will soon succeed in deciding the question raised here."[3]

The experimental physicist Jean Perrin and his colleagues met this challenge in a way that surprised Einstein, who wrote Perrin that "I had believed it to be impossible to investigate Brownian motion so precisely." The simplest approach is to watch particles jiggling under the microscope and calculate their average motion over time. As the notes explain, this gives results for Avogadro's number of the order of 10^{23}, which already is striking because such microscopic particles seem totally different from sugar solutions. Further accuracy requires determining the radius of the particle. To do so, Perrin used gamboge (a kind of soft rubber) that could be painstakingly rubbed by hand into tiny globules. Using a centrifuge, he could select globules of a definite and uniform size. Other exacting procedures allow measurement of their radius. Sometimes protozoa would invade Perrin's rubber-droplet world and he would have to wait days for them to die.[4]

Gazing for hours through the microscope, Perrin calculated that the globules, like van t' Hoff's solutions, act exactly as if they make up an ideal gas, even though the largest globules are just visible under a magnifying glass. He then performed the experiments Einstein had suggested, watching the globules execute their random dance (figure 8.1). Perrin noted the resemblance of their jagged motion to the curves that interested contemporary mathematicians, curves so discontinuous that they do not have a tangent at every point. These jagged paths are the mark of the atomic reality underneath. They are examples of the "drunkard's walk," a random process found in many statistical problems.

Figure 8.1
Jean Perrin's illustration of a typical path of a particle executing Brownian motion, from his book *Atoms*.

Perrin then made a scatter plot of the observed distances traveled by one of his globules after one minute (figure 8.2), which could also describe many other examples of random walks. The plot shows a characteristic grouping, denser near the starting point at the center, from which Perrin calculated an average (root mean square) that can be put into Einstein's equation. This brief summary does not do justice to the excruciating, tedious effort that Perrin devoted to improving this procedure to a high level of precision that gave N_A as about 7×10^{23}.

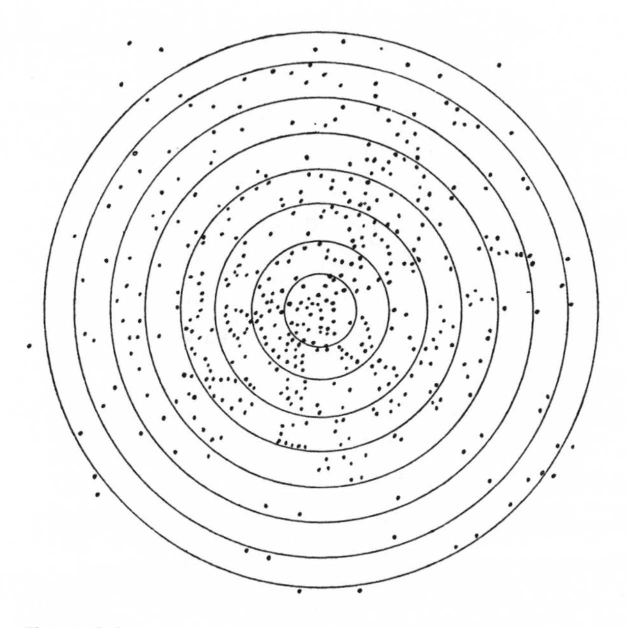

Figure 8.2
Perrin's plot of the results of a set of experiments. Each dot shows how far the observed particle has moved from its starting point (the center of the target pattern) after one minute.

Radically different approaches also gave comparable values. By measuring the number of particles given off in radioactive decay, especially their fluctuations, one can determine the number of atoms emitting them. This led to very accurate value of Avogadro's number, about 6.2×10^{23}. In 1907, Einstein proposed measuring the voltage fluctuations in an electric circuit. As with Brownian motion, the random fluctuations bear the mark of the atoms that underly them and thus yield another value for N_A.

By 1914, Perrin compiled all the methods then known (including two others I will discuss shortly), which gave values of Avogadro's number converging somewhere between 6 and 7

$\times 10^{23}$. Even in 1910, when the data were less refined, he had concluded that "I think it impossible that a mind, free from all preconception, can reflect upon this extreme diversity of the phenomena which thus converge to the same result, without experiencing a very strong impression, and I think that it will henceforth be difficult to defend by rational arguments a hostile attitude to molecular hypotheses." In 1908 Ostwald conceded that Perrin's work constituted "an experimental proof" of the atomic theory. By 1914, Perrin concluded his classic book *Atoms* by stating that "the atomic theory has triumphed. Its opponents, which until recently were numerous, have been convinced and have abandoned one after the other the skeptical position that was for a long time legitimate and even useful."[5]

In this triumph, results about skylight found a surprising connection with a seemingly unrelated phenomenon. In 1869, Thomas Andrews had been studying carbon dioxide near its "critical point," the highest temperature at which it can make the transition from gas to liquid, no matter what the pressure. Andrews noticed that, very near to its critical temperature (31°C), carbon dioxide suddenly became strikingly cloudy in appearance and slightly blue. This phenomenon was eventually called "critical opalescence" because of its resemblance to the milky appearance of an opal stone. Figure 8.3 shows carbon dioxide under high pressure cooling down past the critical temperature, reading from left to right. The image furthest to the left (a) shows the gas well above the critical temperature. In the second image (b), the milky opalescence appears about one degree above the critical temperature, disappearing either above or below that temperature.[6]

This puzzling phenomenon indicated some special state near the critical point, which atomic theory needed to explain. The

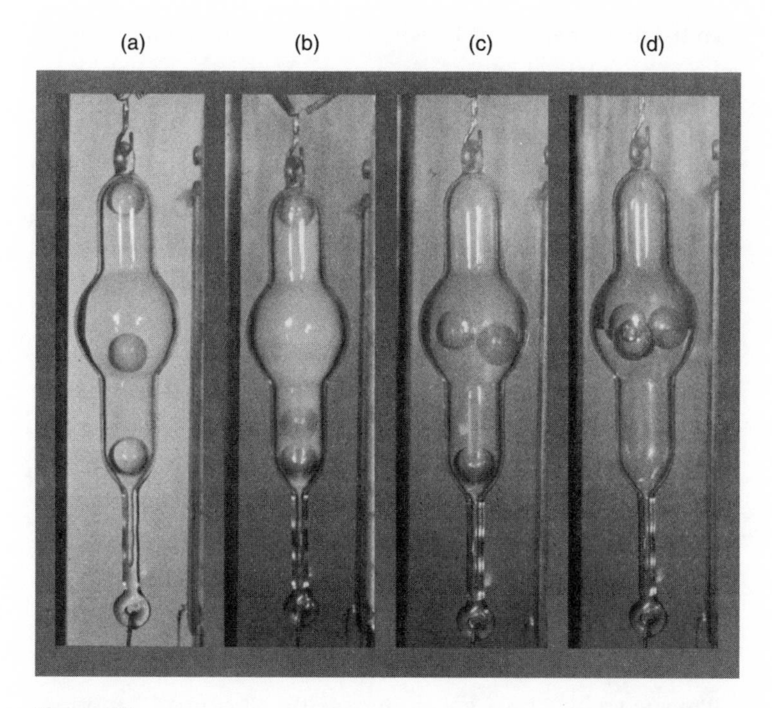

Figure 8.3

Critical opalescence in carbon dioxide near the critical density. (a) The temperature T is greater than the critical temperature, T_c. (b) T is just above T_c. Note the milky opalescence. (c) T is just below T_c and the opalescence has ceased. (d) T is much lower than T_c. The three glass balls inside the tube have densities greater, equal, and less than the critical density, respectively, and their movements show the changing density of the carbon dioxide as it passes the critical point. (Courtesy Jan Sengers. Reproduced with permission from *Chemical & Engineering News*, June 10, 1968, 46[25], p. 105. Copyright 1968 American Chemical Society.)

seminal figure here is the Polish physicist Marian Smoluchowski, who first explained critical opalescence as an atomic manifestation *par excellence*. In 1907, he argued that, as the substance approached its critical point, there would be fluctuations in its density because the constituent molecules were on the verge of regrouping into a new spatial configuration. At that point, a dispersed colloid of tiny liquid globules suddenly pervades the gas. One might compare this to the larger and smaller bunches of concertgoers milling about, returning to their seats after intermission, their state changing from a "gaseous" crowd in the lobby to a "solid" assembly back in their seats.[7]

Indeed, fluctuations are a quintessentially atomic phenomenon, for if atoms underlie all appearances, then the random dancing of those atoms will result in fluctuations in many observable quantities. For example, Brownian motion is a particularly simple example of such fluctuations in position, whose statistical laws Smoluchowski discovered independently in 1906, almost simultaneously with Einstein's work in 1905.

Prepared by his encounter with Brownian motion, Smoluchowski was ready to recognize its underlying similarity with critical opalescence. The difference, however, is that, whereas Brownian motion is continual, critical opalescence emerges in a narrow range of temperature. Smoluchowski showed that the special conditions very near that point greatly amplify the observable fluctuations, to the point that they cause a cloudy opalescence visible even to the naked eye. He concluded that

> These agglomerations and rarefactions must give rise to corresponding local density fluctuations of the index of refraction from its mean value and thus the coarse-grainedness of the substance must reveal itself by Tyndall's phenomenon, with a very pronounced maximal value at the critical point. In this way, the critical opalescence explains itself very

simply as a result of a phenomenon the existence of which cannot be denied by anybody accepting the principles of kinetic theory.[8]

"Coarse-grainedness" refers to the atomic nature of the substance, and "Tyndall's phenomenon" is just the preferential scattering of shorter wavelength (bluer) light. Smoluchowski was fascinated by Tyndall's attempt to recreate the sky's color in a glass tube. Furthermore, Smoluchowski thought the bluish color of critical opalescence was essentially the same as the sky's blue.

What could this mean? Recall that Tyndall thought his scattering was due to very small particles suspended in the gas, since he could find no evidence of scattering in clean, pure air by itself. Rayleigh implicitly considered the scattering molecules to be fixed, whereas Smoluchowski emphasized their ceaseless random motion. Smoluchowski thought that fluctuations in density of those molecules also scatter light, in essentially the same way the molecules do. We will have to examine whether the *real* cause of the sky's blue is molecules or fluctuations in density.

Here, let us pause to consider the role interference might play between these many different scatterers, given that each is excited by the incoming light and emits its own light wave in response, as Huygens first realized. To what extent would all those light waves interfere, destructively or constructively? If the separate scatterers respond in synchrony to the incoming light, we would expect to observe interference, just as Young did from his slits, whose alignment correlated the waves they emitted. Consider *coherent* scattering of light from a solid or liquid in which the molecules are sufficiently densely packed that they respond to incoming light not independently but in a highly correlated way. This occurs when the average distance between molecules is much lower than the wavelength of the incident light (400–700 nm for visible light). For instance, the molecules in glass are separated by a few nanometers and thus satisfy this requirement. In this case, an incoming beam of visible light interferes constructively along its incident direction, manifest as its "index of refraction," but interferes destructively away from that direction Thus, there is no appreciable scattering away from the incident direction, so that glass looks transparent. This also applies to air at sea level, for which the intermolecular separation is roughly 300 nm, lower than the wavelength of visible light.[9] But if X-rays of wavelength under a nanometer illuminate a solid, then scattering can occur, as Max von Laue discovered in 1912, determined by the solid's atomic structure.

In contrast to air molecules at sea level, those in the upper atmosphere move randomly and independently; their average separation is greater than the wavelength of light. Thus, they scatter incoming visible light *incoherently*, meaning that there is no correlation between the light waves emitted by different atoms. Because of this, they are not in phase with each other and so any possible interference is essentially destroyed. Thus, the net effect of all of them is just the sum of the scattering caused by each separately, so that the blue sky emerges as the result of these independent scatterings. Notice that this happens only if an appreciable part of the atmosphere is sufficiently attenuated that the molecules scatter incoherently. The implication is that, were the whole atmosphere somehow held at standard sea-level pressure, the sky might be clear as glass, not blue. We will return to this curious possibility in chapter 10.

Though he suspected a connection between opalescence and the blue sky, Smoluchowski did not try to quantify it, realizing that this would involve extensive modifications of Rayleigh's treatment. That was left for Einstein to do in 1910.[10] He had been struck by critical opalescence since his student days; in the margins of one of his university notebooks, next to his professor's description of critical phenomena, Einstein jotted down "obscure point." A true student, his attention was drawn to what his teachers found puzzling.

When he read Smoluchowski, Einstein was ready to make detailed calculations, prepared by his previous studies of Brownian motion and other phenomena that involve fluctuations. He had developed powerful and elegant mathematical techniques to deal with such problems, of which critical opalescence proved to be another. His result was a formula very similar to Rayleigh's scattering law. Einstein calculated that the scattered intensity of light from an opalescent fluid is inversely proportional to the wavelength to the fourth power and to N_A, also involving factors depending on temperature, pressure, and the index of refraction. As Smoluchowski suspected, the blue sky and an opalescent fluid follow the same basic law, including the same tell-tale polarization of the scattered light.

Though Einstein did not complete the calculation of Avogadro's number based on opalescence, Perrin showed that it yields an accurate value, 7.5×10^{23}. It should be kept in mind that this result, as well as the other values listed, all have errors of about 15 percent and thus are all consistent with the currently accepted number, 6.02×10^{23}. Einstein's 1910 paper on opalescence was his farewell to Zürich before moving to Prague to begin his long struggle to complete general relativity.

In this paper Einstein notes that "it is not surprising that our opalescent light shares this property with the opalescent light produced by suspended particles that are small compared with the wavelength of the light. After all, both cases involve irregular disturbances of the homogeneity of the irradiated substance, the locations of which are rapidly changing."[11] The fluctuations in density and irregular motion of atoms are not incidental, as Rayleigh seems to have assumed; Einstein shows that they are the most ubiquitous and accessible sign of molecular reality. Ironically, the quantitative results of Einstein's paper were (as he himself pointed out) not only true close to the critical point, but valid far away from it. Though leaving further investigation into the mysteries of critical opalescence for his successors, Einstein thus established a general connection between ubiquitous fluctuation phenomena (manifest near the critical point as opalescence) and the blue sky.

Even Smoluchowski had some difficulty with the subtleties of this connection. In 1911, he published a paper that attributed the sky's color to *two* causes: Rayleigh scattering by the molecules of the air and what he thought was the altogether different scattering from fluctuations that he and Einstein had investigated. Einstein immediately wrote him to say that "a 'molecular opalescence' *in addition* to the fluctuation opalescence does not exist." Here the residual influence of the old concept of static molecules comes forward. Smoluchowski evidently thought that fluctuations in density could be separated from molecules considered as static beings; Einstein insisted that the fluctuations *were* molecules, so to speak, because the two are indissolubly linked. Smoluchowski wrote back to tell Einstein "You are completely right."[12] Despite this, the false claim that the sky's blue is "really" caused by fluctuations, not molecules, has been repeated innumerable times since Einstein refuted it in 1911. But a molecule can hit a surface or scatter light, whereas a fluctuation cannot; one is physically real, the other a statistical concept.[13]

The emergent quantum theory was shaped by such considerations about statistics and fluctuations. During the nineteenth century, it became increasingly clear that all bodies emit radiant energy according to their temperature, though that energy takes the form of visible light only at high temperatures (such as when the body becomes red hot or white hot). In 1900, Max Planck first proposed a formula for the spectrum of the radiant energy emitted at a given temperature by a "blackbody." In his treatment, a blackbody is essentially an oven whose walls are sufficiently dense and thick to be opaque to the radiation it emits and with which it is in equilibrium. (The spectrum of an ordinary incandescent light bulb is a good approximation to a blackbody spectrum.) This universal spectrum depended on a new fundamental constant, now named Planck's constant, deeply significant for quantum physics.[14]

In 1905, Einstein noted that in the limit of long wavelengths, Planck's formula also determines Avogadro's number. This is especially striking because the blackbody can be made of any material. The radiant energy it emits seems at first to have nothing to do with atoms, if one thinks of Maxwell's understanding of light as electromagnetic waves. Yet the light from the blackbody is "atomic" because it can have only discrete amounts of energy, which became clear only after Einstein's work in 1905. Ultimately, the atoms emit and absorb radiant energy in equilibrium. As a result, Avogadro's number is manifest both in the characteristics of the light and of the atoms.

As we earlier roughly determined Avogadro's number from skylight, we can also estimate Planck's constant from sunlight because the sun itself is approximately a blackbody. The notes detail how we can compare the sun's warmth on our skin to that of a light bulb of known wattage and, from that, calculate that the sun's effective temperature is about 5,800 degrees

Kelvin. Then Planck's formula connects his constant with this temperature, the wavelength of visible light, and two other fundamental constants (the speed of light and Boltzmann's constant). The result gives a range of values for Planck's constant of the right order of magnitude. Gazing into the sky allows us to calculate the quantum constant, not only Avogadro's "classical" number.[15] But as Einstein had noted, these basic numbers must finally have a consistent connection.

Like so many others in our story, Smoluchowski had a passion for mountaineering. As if going on a final expedition, he returned to the quest for sky blue in 1916, this time as an experimenter. He wanted to use opalescence to recreate the sky in a bottle, though in a different way from Cabannes's direct scattering of light in clean air. Smoluchowski planned to measure the scattering of light from opalescent gas mixtures to test Rayleigh's theory, and his preliminary experimental results were promising, though slowed by World War I. His 1916 paper announced the continuation of his experiments, which succeeded in producing blue light, especially using hydrogen gas and ethyl chloride. Unfortunately, Smoluchowski's work was cut short by his death soon thereafter. Einstein mourned the passing of "a noble and subtle human being."[16]

Their contemporaries were engaged in other attempts to capture the sky. The romantic sensibility had exalted the ceaseless drama of the clouds. Goethe was impressed by the cloud studies of Luke Howard, a British chemist who introduced the modern nomenclature for cloud types. Especially in Britain, the fascination with clouds was growing, depicted with a new level of sensitivity by John Constable, in whose paintings clouds became a central subject in their own right, not merely as atmospheric details. For a full year after the explosion of

Krakatau (Krakatoa) in 1883, the sky was filled with extraordinary phenomena, such as blue suns (from the dust-filled air) and exceptionally colored sunsets.[17]

You can see a small but genuine cloud form for an instant just after you open a bottle of any carbonated drink. The quest to bottle clouds involved many scientists and led in surprising directions. Among them, consider C. T. R. Wilson, who from his youth devoted himself to photographing the clouds. He continued this artistic hobby when in 1894 he joined the new weather station on Ben Nevis, the highest peak in the British isles: "The wonderful optical phenomena when the sun shone on the clouds surrounding the hilltop, and especially the coloured rings surrounding the sun (coronas) or surrounding the shadow cast by the hill-top or observer on mist or clouds (glories), greatly excited my interest and made me wish to imitate them in the laboratory." The next year he built his first "cloud chamber" to mimic these wonders, in which dry ice and alcohol recreate the supercooled conditions of the upper atmosphere.[18]

Unlike other experimenters with such devices, Wilson's interest turned to ways in which his chamber could detect the new phenomena of X-rays and radioactivity. Influenced by J. J. Thomson, the discoverer of the electron, Wilson became increasingly involved with this new world of microphysics, though he worried that he had lost touch with his initial plans to understand the weather. His new direction led him to filter out the very dust he knew enabled the formation of ordinary clouds.

Despite Wilson's fears, the cloud chamber became a crucial detector of new kinds of celestial phenomena. In the chamber's miniature sky, charged particles leave behind tiny clouds whose curvature and density reveal their identity in tracks analogous to those formed by jet planes. The cloud chamber was an essential tool in the study of cosmic rays, whose mysteries opened the way to the vast realm of fundamental particles; in it were discovered the muon, the pion, and the positron, the first known antiparticle (figure 8.4). Modern particle physics would have been unimaginable without it; in this bottle you can see wraithlike clouds appearing in the wake of invisible particles.[19]

As science reinterpreted the sky and its blueness, art both extended and questioned its place in the romantic sensibility. In 1910–1911, Wassily Kandinsky and Franz Marc founded an almanac they called *The Blue Rider* because "we loved blue, Marc horses; I riders. So the name invented itself." Blue is also central to Kandinsky's manifesto *Concerning the Spiritual in Art* (1911). Though his avant-garde painting seems far from romantic art, his exaltation of blue was no less fervent. Goethe's color theory remained fascinating to Kandinsky, who also considered blue a "cool" and distant color as opposed to yellow. Kandinsky thought of blue as "the *male* principle, sharp and spiritual," as opposed to its romantic characterization as female. He remarked that "the deeper the blue becomes, the more strongly it calls man towards the infinite. . . . Blue is the typical heavenly color. Blue unfolds in its lowest depths the element of tranquility. As it deepens towards black, it assumes overtones of superhuman sorrow." Kandinsky saw in "the high, pale-blue sky" something "remote and impersonal."[20]

In the same vein, Tolstoy's Prince Andrei, lying wounded on the battlefield, gazed with wonder at "the lofty sky": "All is vanity, all falsehood, except that infinite sky. There is nothing, nothing, but that. But even it does not exist, there is nothing but quiet and peace. Thank God!" On another occasion, a Russian officer observes that "whatever we may say about the soul going to the sky . . . we know there is no sky but only an

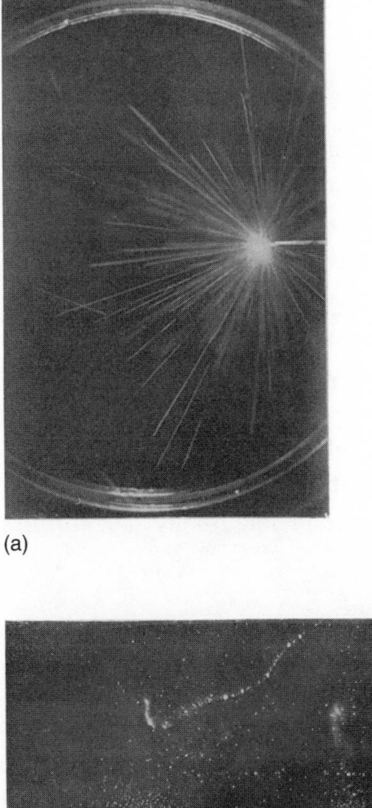

(a)

(b)

Figure 8.4
C. T. R. Wilson's cloud chamber photographs (1912), showing (a) alpha particle (helium nucleus) tracks; (b) both alpha and beta (electron) tracks. The much greater density of the alpha tracks indicates their greater charge and much greater mass, compared to the electron.

atmosphere." And later still, Andrei's friend Pierre says "we must live, we must love, and we must believe that we live not only today on this scrap of earth, but have lived and shall live forever, there, in the Whole," pointing to the sky.[21]

For these artists, the sky was transcendence made visible. But for others, the sky was just an atmosphere, coldly impassive. As Emma Bovary's father rode home, he could not believe that she was dead, for "the sky was blue." That immense vault seems not to regard or reflect our sufferings. It may be "ironically and cruelly blue," as it was to the swan in Baudelaire's poem "Le Cygne," about a wild bird lost in the stony alleys of the modern city. The swan "raises its convulsive neck and avid head as if it reproached God," but the sky is silent. Stephane Mallarmé wrote in 1864 that "the sky is dead." In his poem "L'Azur," the blue sky is the unattainable beauty that tortures the poet:

The serene irony of the eternal azure,
Lazily beautiful like the flowers,
Crushes the impotent poet who curses his genius
Across a sterile desert of Sorrows.[22]

That same ironic sky loomed over the grim trenches of World War I. As the writer Max Plowman noted:

Was it Ruskin who said that the upper and more glorious half of Nature's pageant goes unseen by the majority of people? . . . Well, the trenches have altered that. Shutting off the landscape, they compel us to observe the sky; and when it is a canopy of blue flecked with white clouds . . . , and when the earth below is a shell-stricken waste, one looks up with delight, recalling perhaps the days when, as a small boy one lay on the garden lawn at home counting the clouds as they passed.[23]

This was a generation educated by Ruskin to search the sky "for the perpetual comfort and exalting of the heart, for the soothing it and purifying it from its dross and dust."[24] But with what mixed and turbulent emotions the dying must have gazed into that overarching blue. Did they share Prince Andrei's mysterious consolation, or did the sky's very beauty magnify their despair?

Even after all the horror, the blue sky retains a haunting fascination, as in Wallace Stevens' "Sunday Morning" (1915), in which a woman dreams that

The sky will be much friendlier than now,
A part of labor and a part of pain,
And next in glory to enduring love,
Not this dividing and indifferent blue.

9 Midnight Blue

The sun has set, the moon is low, and we contemplate a clear sky. Compared to the bright day, we now reconsider both the sky's color and its faint luminosity. It certainly looks "as black as night" to most observers, but it is not truly black. The deep sky is sufficiently dim that the color receptors of our eyes do not register it in the same way that they register the daytime hue of the sky, even though the rods of night vision are more sensitive to blue wavelengths than are the cones of day vision.

But in a really dark place with no moon, the sky glows faintly, enough that you can see your outstretched hand against it and even count your fingers. Part of this glow is a kind of aurora. Particles streaming out from the sun excite nitrogen and oxygen in our atmosphere, which then glow a faint yellow. This effect waxes or wanes depending on the varying flux of particles. The younger Lord Rayleigh recorded a series of observations of the color of the night sky in 1920 and 1923, both at home and in Sicily. His conclusion was that about 10 percent of the night skylight was auroral, based on its observed color. In addition, the stream of solar electromagnetic radiation excites air atoms and molecules to chemiluminescence, giving a continuous airglow. Other, fainter sources of background skylight include the reddish hues of glowing sodium (from sea salt or meteoritic dust), zodiacal light (from dust in the solar system), and even fainter light from distant stars and galaxies.

When the moon is up, its light scatters through the atmosphere. This molecular scattering makes the night sky *blue*, though at such low light levels that the human eye cannot really register it. Here the camera gives decisive evidence. During the day, take a color photograph of a patch of sky. Then, at night when the moon is not near the field of view, aim the camera toward the patch and take a very long exposure using a tripod. When both photographs are developed, they show the same blue. (The notes give suggestions about exposure and settings.)[1]

Here, we are concerned with the appearance of the sky itself, apart from the stars visible at night. This may seem superfluous to add, were it not for a myth (dating back to Aristotle but still current) that stars can be seen during daylight from the bottoms of deep mines, wells, or through long chimneys. Alexander von Humboldt inquired repeatedly during his activities in mining, but never found a miner who had seen this. Though in daylight you can sometimes see Venus if you know exactly where to look, it would be difficult to see stars up a mine shaft or chimney even in the middle of the night, because the narrow field of view would be unlikely to include a star. During the day, the brightness of the sky is too great, the sensitivity of the eye too weak.[2]

But there remains an important problem about the relative brightness of the night sky that leads to a surprising insight. Several times already in our story we have encountered what we have called the night sky puzzle: If the earth were surrounded by an unlimited number of stars distributed throughout space, the night should be as bright as the day, and both should be as bright as if the sun filled the whole sky.[3]

Kepler took the darkness of night to show there could not be an infinite number of suns spread through space, which he considered an abhorrent thought, as venturesome as he was. Yet the puzzle does not require the stars to be infinite in number. Consider a forest that is finite but sufficiently dense that there is a *background limit*, a distance past which you cannot see because your line of sight will have met some tree. If there is a background limit in the cosmos, every line of sight will eventually meet a star and the night sky will be uniformly bright even in a finite universe. Contrary to Kepler, infinity is not the issue.

Since his time, the night sky puzzle has been rediscovered many times, its history forgotten. It is often referred to as "Olbers's paradox," after the early-nineteenth-century astronomer who stated it yet again. Many ingenious solutions were offered, which Edward R. Harrison details in his classic book, *Darkness at Night*. Let me recount a few to prepare the ground for the new insight I would like to add.

In 1744, Jean-Phillip Loys de Chéseaux calculated that a star-filled sky would be ninety thousand times brighter than the sun, assuming that each star was as bright as the sun itself. This would require 10^{46} stars distributed uniformly, for which the background limit is about 6×10^{15} light-years, a million times the size of the observable universe as currently known. This distance is so vast that Chéseaux felt there must be some absorption throughout space, however slight, that would consume the incoming light from the distant stars and render the night dark. But in 1848 John Herschel responded that any absorbing interstellar medium would eventually radiate "from every point at every instant as much heat as it receives." If so, absorption does not solve the puzzle, but only delays it. Herschel's argument itself took about a century to sink in and required much clarification of the nature of the necessary equilibrium between absorption and emission in such an understanding of the universe.

Reviewing Humboldt's *Cosmos*, which mentioned the puzzle, Herschel suggested that the stars were not distributed uniformly, but clustered in hierarchies: a cluster of stars is part of a much larger and less dense cluster of clusters, and so on up the hierarchy. If the successive levels of structure have the right law of decreasing density, the puzzle is solved. But this solution requires a very particular distribution of stars and also an ever-expanding hierarchy extending to infinity. If the hierarchy stops at some finite level, the paradox will return, given a sufficiently large number of stars that we are in the same situation we began with.

An even more modern approach invoked the expansion of the universe, in which distant galaxies recede from each other according to a universal law discovered by Edwin Hubble: the velocity of apparent recession is directly proportional to the distance between the galaxies. The light received by such mutually receding luminous objects would appear shifted to the red by the Doppler effect. Planck showed that photons of red wavelengths have less energy than blue. Thus, the photons from ever more distant galaxies would be so red-shifted and miniscule in energy that they would not contribute appreciably to the brightness of the night sky. This ingenious theory implies that the night sky is dark because the universe is expanding!

Yet cosmic expansion does not really solve the night sky puzzle, for which a much simpler explanation suffices, first published by Lord Kelvin in 1901. The missing factor is that stars have a finite lifetime and light has a finite velocity, so that, past a certain distance, the light has not yet reached us, and moreover the stars cannot live long enough for the night sky to become bright. As Harrison discovered, Edgar Allen Poe had uncannily anticipated this solution in an ecstatic cosmological "prose poem" he called *Eureka* (1848), which he dedicated to Alexander von Humboldt. Poe writes:

> Were the succession of stars endless, then the background of the sky would present us an uniform luminosity, like that displayed by the Galaxy—*since there could be absolutely no point, in all that background, at which would not exist a star.* The only mode, therefore, in which, under such a state of affairs, we could comprehend the *voids* which our telescopes find in innumerable directions, would be by supposing the distance of the invisible background so immense that no ray from it has yet been able to reach us at all.[4]

Here, as with Ruskin twenty years afterward, a visionary was able to discern a conclusion vindicated only much later by sober science. Far more often, though, searching analysis disproves the beautiful surmise. Basic truths eventually make themselves felt. For instance, long after Kelvin, Harrison showed that there is simply not enough energy in the universe to create a bright night sky. Neither considerations of cosmic expansion nor those of the curved spacetime required by general relativity can prevail over that basic lack of energy. Though this solution seems definitive, the night sky puzzle has still another side that remains to be examined.

The night sky nevertheless has a certain faint luminosity. Here, we are not considering the light from bodies in our solar system, which includes the zodiacal light that so intrigued Humboldt as well as other, fainter near-earth phenomena: light pollution from the ground, airglow (luminescence in the atmosphere), and gegenschein (similar to the zodiacal light, but seen opposite to the sun in the sky). Nor are we interested in the light from nearby stars, but rather the generalized luminosity that still shines between those stars. Though there are patches in the night sky that seem blank and dark, the telescope reveals further stars in every such patch. We cannot find a finite patch of sky that does not contain a measurable brightness.[5]

Granted, then, that this background brightness is small, why is it as large as it is, rather than much more or less? We can use Kelvin's argument, run backwards, to reveal the connection between the observed brightness and the density of luminous stars.[6] Kelvin shared the cosmology of his times, which assumed a single island galaxy surrounded by empty space. This was not because the notion of many galaxies had not been raised. As mentioned earlier, Wright, Kant, and Lambert all had suggested the many-island view of the universe. William Herschel, the foremost astronomer of the eighteenth century and the father of John Herschel, had for a time championed this view, but toward the end of his life he changed his mind and went back to the one-island universe (figure 9.1). This remained the standard view through the nineteenth and until the early twentieth century. But the controversy reemerged in the so-called Great Debate of 1920 between the partisans of the two competing

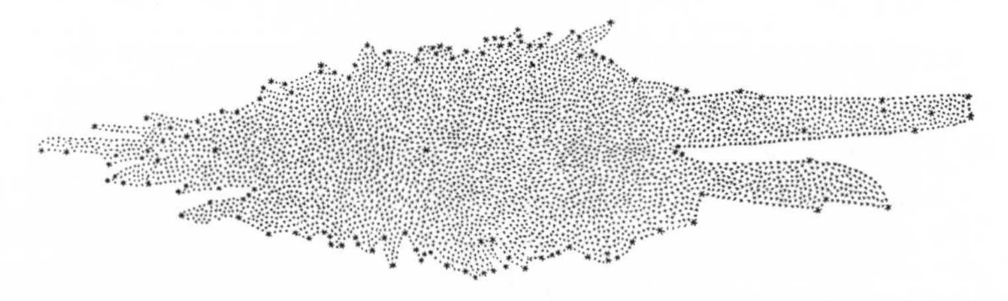

Figure 9.1
William Herschel's diagram of the one-island universe, with the sun at the center. The varying thickness reflects the different densities of stars in those directions, as seen from our solar system.

worldviews. Such eminent astronomers as Harlow Shapley still defended the one-island theory until 1924, when the weight of Hubble's observational evidence became overwhelming on the side of the many-island universe.

Kelvin himself seemed quite sure about the one-island universe of William Herschel, which contained a single "galaxy or region of stars," with the sun at the center, surrounded by a "region of nebulae" on all sides (figure 9.2). In the one-island

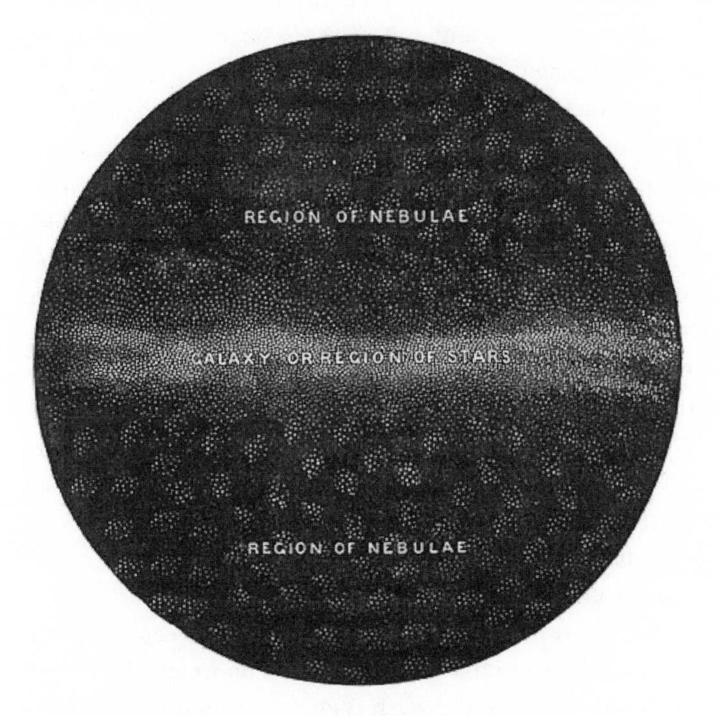

Figure 9.2
Kelvin's one-island universe, as depicted by Simon Newcomb (in his book *Popular Astronomy*, 1878).

cosmos, the single galaxy has about one billion stars spread through 3,300 light-years. This was the well-accepted consensus of contemporary astronomers. Kelvin assumed, for simplicity, that all the stars have the same intrinsic luminosity as the sun. Using these numbers, he calculated that only one-trillionth of the sky is covered with stars. To make the night sky dense with stars, and hence bright, he calculated that the universe would need to be filled with stars to a background limit of 3,300 trillion light-years, vastly larger than the universe he knew. But even this is not sufficient unless the stars shine indefinitely, which Kelvin knew was not the case.

Kelvin himself had done important work on the problem of the lifetime of stars. At first, he had thought that asteroids falling into the sun were the source of its heat. Later, he realized this was insufficient and proposed that the sun derived its energy from slow gravitational contraction, from which he derived a lifetime of twenty million to one hundred million years. Though far longer than the time span traditionally allowed by biblical chronology, Kelvin's lifetime seemed far too short in the opinion of a number of geologists. From studies of the observed rate of erosion, they arrived at much longer times than Kelvin, in the region of five hundred million to one billion years. In fact, Kelvin was wrong, but such was his scientific prestige that his pronouncements about the age of the earth had a chilling effect on geology.[7]

The matter was resolved only in the twentieth century with the advent of radioactive dating, which confirmed the suspicions of the geologists: the earth is about five billion years old, far older than Kelvin's one hundred million years. In 1928, Hans Bethe and others used quantum theory to show how thermonuclear processes could be the real source of stellar energy, giving stellar lifetimes much closer to the geological times. Kelvin had died long before, in 1907. But let us go back to his calculations, in which a surprising insight lies hidden.

Kelvin's equations remain quite correct, even today. Once he had used them to establish that the stars do not live long enough to create the paradoxical brightness of the night sky, he did not examine them further. If we go back to them, though, we can derive a simple relation between the number of stars per unit volume and their lifetime, based on an estimate of the actual brightness of the night sky. Kelvin could easily have found this relation (discussed further in note 6), which is a simple rewriting of his own equations.

Now imagine that Kelvin had decided to use in this relation what he knew of the one-island cosmos. From his data, he would then have concluded that the lifetime of a star is only about three hundred years! Even if he pushed his data as far as they would go, he could only have gotten a lifetime of three thousand years, which contradicts the fifty million to one hundred million years he had calculated. Turning this the other way around, based on what he knew of the number of stars and their lifetimes, he would have concluded that the night sky should be at least ten thousand times brighter than it is.

Thus, following this imaginary train of thought in 1901, Kelvin could have concluded from his own theory that the cosmology he had assumed was radically in contradiction with observation. He had declared in 1884 that "there are no paradoxes in science," only our misunderstandings.[8] Accordingly, he would have felt obliged to remedy this glaring contradiction by removed one of its presuppositions. Taking the brightness of the night sky as an observational given, he would have had to abandon either his assumption about the number of stars or his assumption about their lifetime, if not both. Let us say that he refused to change his estimate of their lifetime, especially since he felt he had such strong grounds to think he was right. Then this same relation would have told him that the number of stars in a unit volume is ten thousand times smaller than what he and everyone else had assumed.

That conclusion could only have meant that the one-island galaxy he assumed was contradictory. He then could easily have found an alternative solution in the many-island universe, in which the stars are now spread over a far vaster space than one small island. Thus, Kelvin in 1901 could have found and solved a great paradox of cosmology and prophesied the many-island universe decades before it was established.

This discovery would have been probably the greatest advance in cosmology that occurred through pure thought. The deduction involves only the general estimate of the brightness of the night sky, certainly not the photographs taken at the Mount Wilson and Palomar Observatories that established that many "nebulae" were really galaxies far distant from our own. In fact, Chéseaux in 1744 or Olbers in 1823 could have made the same deduction since they already knew about the importance of the background limit. Olbers, in particular, knew the rudiments of the dynamics of stars and thus could have derived the basic relation that we have been discussing. Yet both of them believed (wrongly) that the night sky was dark because of absorption and hence neither had any incentive to pursue the matter further.

For all his greatness, Kelvin too had his limitations. Once having solved the "paradox," he evidently did not think to keep probing it. His disbelief in paradoxes also led him to think that they were mere misunderstandings that could be scotched, rather than the possible starting points of new insights. Our imaginary story is also a cautionary tale about the unspoken power of our assumptions and how difficult it is to see beyond them. After all, we have imagined all this with the power of hindsight, though, to be sure, nothing stood in Kelvin's way other than his presuppositions. When our descendants look back on us, at which of our preconceptions will they shake their heads, wondering at our blindness?

Yet our imaginary history does not merely leave us musing over what might have been. Gazing at the night sky, admiring its midnight blue, we now realize that its brightness is telling us something deep. For instance, we can see by naked eye the neighboring galaxy in Andromeda, so large and bright in a dark sky that ship captains used to think it was a comet. We know that it contains about one hundred billion stars and is about three million light-years away from us.

Let us now take these numbers as typical of the whole universe and make a rough calculation of our own. Let us assume that all galaxies have the same number of sunlike stars and are spaced evenly at the same distance as we are from Andromeda. If so, we conclude that the lifetime of stars should be about one billion years.

Thus, making crude but plausible assumptions based on what we see at night, we have deduced that the universe should be at least one billion years old, not so far from the currently accepted values of roughly fourteen billion years, based on far more accurate data. There is something intriguing about such a deduction from the faint brightness of the night sky, especially since we see only the visible wavelengths.

Of course, there are many other wavelengths of radiation in the sky that our eyes do not register, down to the microwave radiation permeating the universe, now understood to be the remnant of the primordial explosion. This microwave background is exactly like the radiation of a blackbody at a temperature of 2.725 degrees Kelvin. It is the result of the primordial explosion that began at a temperature over ten billion degrees and then expanded to the present radius of the universe, about fourteen billion light-years. Because of it, even the farthest reaches of intergalactic space are illuminated by this radiation and are not altogether dark.

Though it is omnipresent, this background radiation is not visible to the human eye. But even if the stars of our galaxy went out, the night would not be completely dark. The faint but visible light from distant galaxies would illuminate the darkest night.

10 The Perfect Blue

So far, we have taken a few steps toward understanding what we actually see in the sky. But each step, however small, is significant because it invites us to understand more deeply and more completely. Each new insight can lead us to reconsider the questions that drew it forth. The history of this unfolding understanding reveals what has come before and what may yet emerge.

Especially, I wonder at the action of our brain in all this. Given a number of pitches of different wavelengths, our ears perceive a chord in which each note is distinct. But given a similar mixture of lights of different wavelengths, our eyes integrate them into a *single* sensation, say of "white."[1] This is only the beginning of the complexities of vision. I have throughout spoken of the sky as blue, and there is nothing wrong with that, because our color words come from our common experience. But recall the violet puzzle that has come up several times: If molecular scattering requires that violet wavelengths are more strongly scattered than blue, why is the sky not violet? Rayleigh, of all people, might have wondered about this, but he did not mention it. In 1901, Kelvin noted that violet wavelengths should be seven times more scattered than red, "which explains the intensely blue or violet colour of the clearest blue sky."[2] If the sky is "blue or violet," there is no violet puzzle.

For Kelvin, these two hues seemed close enough that the sky can partake of both. Here enters an element of judgment in color perception. For instance, Newton included indigo in the spectrum in order to have a color corresponding to the semitone E–F in the musical scale (see figure 3.4), though he admitted that the spectral colors "could perhaps be consitituted somewhat differently."[3] Others may have applied the term "violet" so broadly that it often merged with blue. Consider, for instance, "violet eyes," a descriptive phrase sometimes found in fiction.[4] Does that really mean the same hue as the flower, or is it a poetic way of naming a deep blue?

Though there were many accounts of the numbers of colors in a rainbow, over time violet came to be part of a standard list of spectral colors. Newton himself tended to group "blue and violet" together as he described the way his prisms broke up the light, yet he distinguished the two and experimented separately with violet.[5] The development of spectroscopy clarified such issues by presenting its data in terms of wavelengths: Those of about 400 nm are termed violet, 445 nm indigo, 475 nm blue. As described in chapter 6, Rayleigh himself made the first spectrum of skylight in 1871, but did not include violet. In 1888, A. Crova ascended Mont Ventoux and made spectral measurements that, compared to his standard light source, showed that skylight "had a great relative intensity in the violet," though his measurements did not include wavelengths lower than 530 nm (blue-green).[6]

Over the next two decades, others confirmed these findings and extended them into the violet.[7] In 1908, Edward Nichols presented a comparison of Rayleigh's theory with several sets of observations (figure 10.1) over the full visible range.[8] All show a clear maximum in the violet, near 400 nm; we will return to the discrepancies between these curves, but they all show the same general shape. These measurements all involved the human eye, but only to compare the brightness of two images at the *same* wavelength: the skylight at a certain wavelength versus that of a standard source, the sun or a special lamp using gas or oil. Because of this, the relative sensitivity of the eye at *different* wavelengths is not a problem, only the skill of the observer in judging relative brightness. But the eye does restrict the range of wavelengths under comparison to the visible range.

Indeed, why shouldn't the same argument that led to the violet puzzle be pushed further: If ultraviolet wavelengths are scattered more than violet or blue, why isn't the sky ultraviolet? Though it was not stated at the time, this puzzle could have been phrased only after Ritter discovered ultraviolet radiation by its action on photosensitive compounds (1801). In fact, early photographic emulsions were sensitive *only* in the ultraviolet and blue, so that they could scarcely get an image of a sunset. Only when plates and film sensitive in the red became available did sunset photography become practicable. Likewise, not until then could photography really record spectra.

Samuel Langley's invention of the bolometer in 1880 first provided a suitable means to measure brightness at wavelengths both within and outside the visible range. This instrument used a delicate balance of electrical resistances to measure the heat conveyed by a narrow range of the spectrum, for radiant energy transfers heat to any surface it illuminates proportional to its brightness.

Langley's work indicated that most of the energy in the solar spectrum is in the infrared, but very little at wavelengths smaller

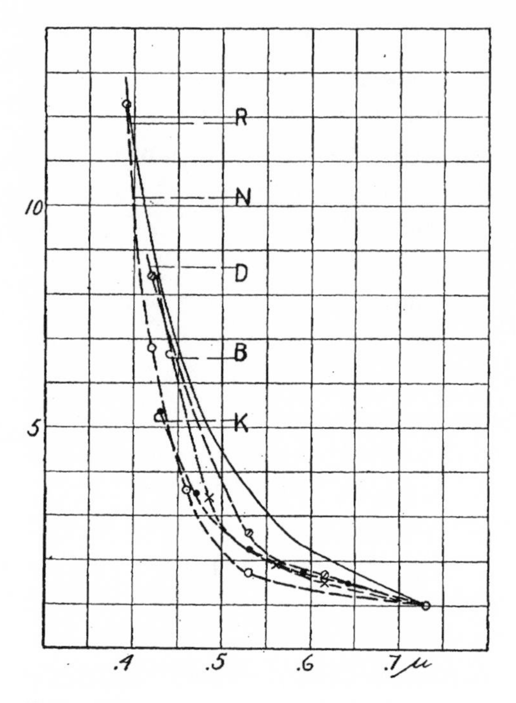

Figure 10.1
Edward Nichols's comparison (1908) of Rayleigh's scattering law (labeled *R*) with different observations of the spectrum of skylight (*N*, *D*, *B*, *K*), showing that actual atmospheric scattering deviates from pure molecular scattering. The vertical axis is an arbitrary measure of brightness; the horizontal axis is measured in micrometers (1 μm = 10^{-6} m = 1000 nm), so that the shortest wavelength shown, 0.4 μm = 400 nm, is violet.

than violet (about 9 percent, according to current values).[9] Thus, the ultraviolet puzzle is resolved because the sun emits so little of its energy in those wavelengths. (Other factors will also emerge shortly.) Modern measurements confirm that, compared to a blackbody, the sun's radiation is noticeably depleted even in the violet, owing to the absorption of those wavelengths by certain elements in the solar atmosphere (figure 10.2). But this does not suffice to explain the violet puzzle. Even despite the sun's poverty in these wavelengths, measurements using modern instruments like the bolometer confirm that the highest peak in the brightness of scattered skylight is in the violet, not blue (figure 10.3).

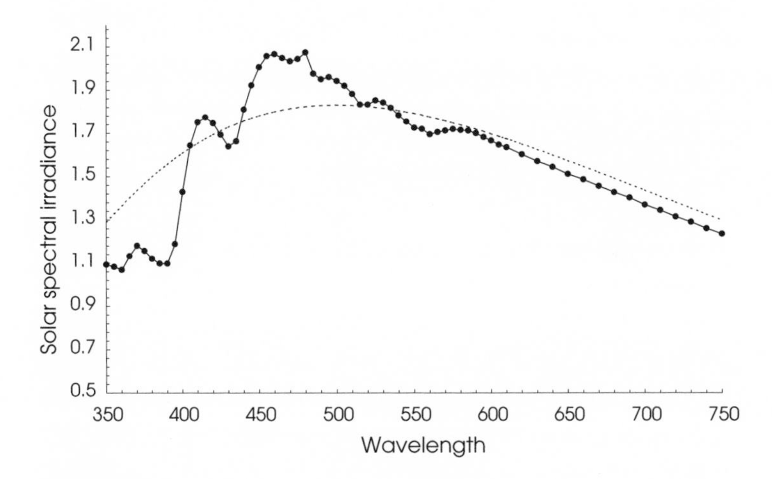

Figure 10.2
Spectrum of sunlight as observed outside the earth's atmosphere (using NASA standard data) showing the similarity of the sun's spectrum (solid line) to that of a blackbody of temperature 5,800 K (dotted line). The vertical axis is solar spectral irradiance (watts per square meter per nanometer); the horizontal axis is wavelength in nanometers. Note the depletion of violet wavelengths near 400 nm.

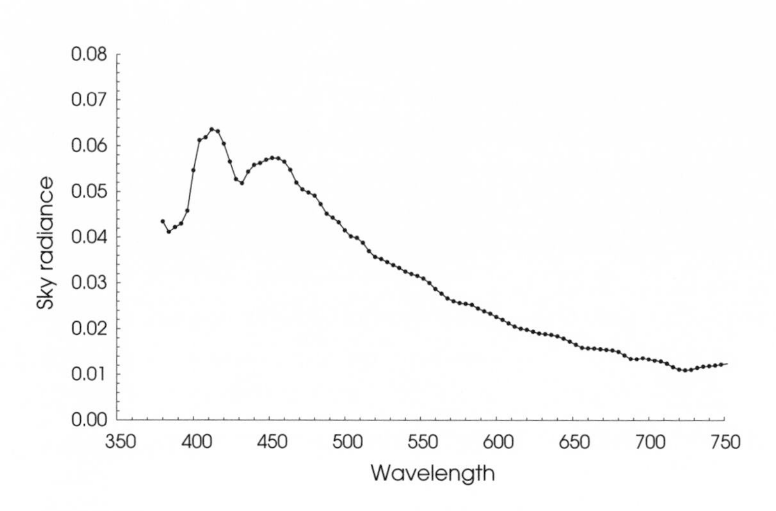

Figure 10.3
Spectrum of skylight as observed at the earth's surface by a spectrophotometer. Note the strong peak in the violet near 400 nm. The vertical axis is the spectral radiance of the sky (watts per square meter per unit solid angle per nanometer); the horizontal axis is wavelength in nanometers. (Courtesy Eugene Clothiaux.)

We do not see the beguiling sight of a violet sky because of the spectral sensitivity of our eyes, which do not register single wavelengths (unlike instruments like spectrophotometers) but rather integrate over all visible wavelengths in a complex way. The peak efficiency of normal daytime vision is about 550 nm (yellow), blue wavelengths are registered with 8 percent of the efficiency of yellow, and violet less than 1 percent (figure 10.4).[10] Our vision simply is far less sensitive to violet, so that it registers a dominant blue hue.

Even so, much remains to be explained about how exactly our brain constructs what we consider sky blue, for color perception is amazingly subtle and surprising. To us, a mixture of pure red

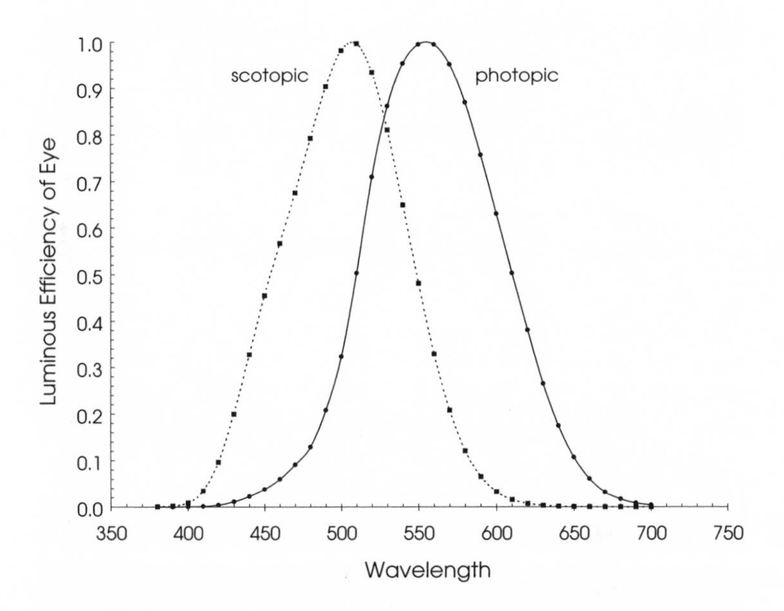

Figure 10.4
Luminous efficiency of the human eye: (a) photopic (daytime) vision (solid line); (b) scotopic (nighttime dark-adapted) vision (dashed line). The vertical axis is the relative efficiency of the eye compared to its relative maximum for each kind of vision; the horizontal axis is wavelength in nanometers. Note the shift of the maximum sensitivity in scotopic vision toward shorter wavelengths, called the Purkinje effect.

and pure green light looks yellow, though the source includes absolutely no yellow wavelengths. If so, the relation between light and vision is very obscure. Indeed, in 1957 Edwin Land demonstrated *full* color photography from only *two* black-and-white images taken through red and green filters and superimposed.[11] Somehow the brain can construct color images even when colors are seemingly absent from the incoming illuminant or photograph. Long before, Newton had even tried to excite

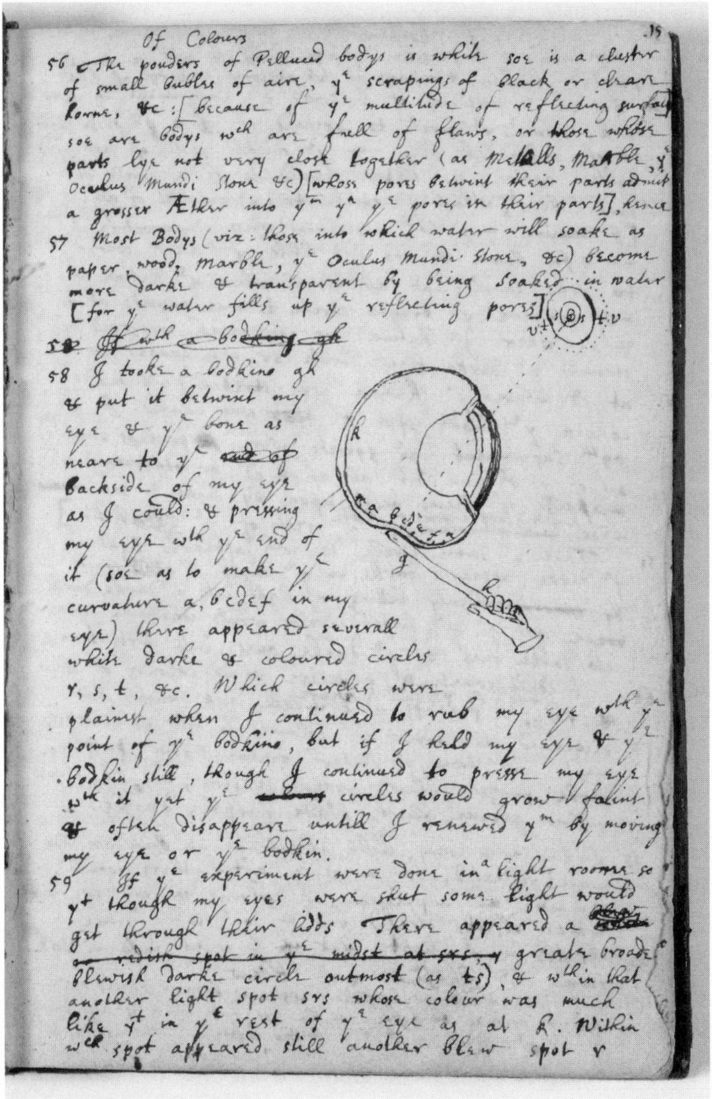

Of Colours .15

56 The pouders of Pellucid bodys is white soe is a cluster of small bubles of aire, ye scrapings of black or cleare horne, &c: [because of ye multitude of reflecting surfaces] soe are Bodys wch are full of flaws, or those whose parts lye not very close together (as Metalls, Marble, ye Oculus Mundi Stone &c) [whose pores betwixt their parts admit a grosser Æther into ym ya ye pores in their parts], hence

57 Most Bodys (viz: those into which water will soake as paper, wood, marble, ye Oculus Mundi Stone, &c) become more darke & transparent by being soaked in water [for ye water fills up ye reflecting pores]

~~58 If with a bodking gh~~

58 I tooke a bodkine gh & put it betwixt my eye & ye bone as neare to ye ~~end of~~ backside of my eye as I could: & pressing my eye wth ye end of it (soe as to make ye curvature a, bcdef in my eye) there appeared severall white darke & coloured circles r, s, t, &c. Which circles were plainest when I continued to rub my eye wth ye point of ye bodkine, but if I held my eye & ye bodkin still, though I continued to presse my eye wth it yet ye ~~[illegible]~~ circles would grow faint & often disappeare untill I renewed ym by moving my eye or ye bodkin.

59 If ye experiment were done in a light roome so yt though my eyes were shut some light would get through their lids There appeared a ~~[illegible]~~ ~~a redish spot in ye midst at srs, a~~ greate broade blewish darke circle outmost (as ts), & wthin that another light spot srs whose colour was much like yt in ye rest of ye eye as at k. Within wch spot appeared still another blew spot r

Figure 10.5

Newton's 1664 description and sketch of his experiment to produce sensations of color by sliding a bodkin behind his eyeball. (By permission of the Syndics of the Cambridge University Library.)

visual sensation by sliding a bodkin (a stiletto or blunt needle) into his own eye socket, working it between the eyeball and the surrounding bone. By pressing the eyeball with the bodkin's tip, he saw "severall white darke & colored circles" that faded when he ceased pressing.[12] One wonders whether he was seeking some new kind of visual sensation or only verifying well-known physiological effects. For instance, a blow to the head, pressure on the eyeball, certain detachments of parts of the eye all can cause color perceptions *without any incoming light at all.* Consider also Oliver Sacks' case history of a painter who became totally colorblind, yet nevertheless could demonstrably distinguish different wavelengths.[13] Such studies indicate that the brain *constructs* color perception in higher centers of its secondary cortex, not simply in the retina. The naive view that identifies color simply with wavelength is untenable; we must always bear this in mind when we consider what our eyes seem to tell us. The sky as we see it always remains in the ultimate bottle: the human brain.

Much is still unknown about human visual processing. For instance, consider the controversy about the evolutionary adaptation of human vision. Note that the peak of the daytime sensitivity of the eye (see figure 10.4) occurs near the wavelength 550 nm, very close to the peak of the solar spectrum (see figure 10.2). Some argue that this is no coincidence; the human visual system has evolved to be as close as possible to the spectral characteristics of the sun. But others respond that, if you consider sensitivity in terms of frequency rather than wavelength (for instance), the match-up between eye and sun is not so good. That is, the question of adaptation looks different depending on how you measure it. If so, the eye may not be so perfectly adapted to the sun. After all, evolution depends on random processes, which may fall short of what we may regard as perfect adaptation. The controversy between these views continues.[14]

Even aside from these complexities of human vision, the mysteries about the sky have not been exhausted. We need to reconsider even the basic connection between the sky's color and molecular scattering. In chapter 7, we concluded that the physical reality of molecules is a *necessary* condition for the sky to be blue. Surprisingly, it is not a *sufficient* condition. In 1984, N. A. Voishvillo and Y. A. Anokhin calculated light scattering in different model atmospheres, each of them purely molecular in composition but characterized by having different numbers of molecules per unit volume, the *number density*. At the number density of the earth's present atmosphere, they recovered Rayleigh's results: The greatest scattering is in the violet (near 400 nm). But if that number density is ten times greater, the curve of skylight brightness shifts so that its peak is in the blue, not the violet. If the number density is thirty times greater, the peak occurs in the orange (near 600 nm).[15] This shows that to have a sky that looks like ours, not only must air molecules be physically real, but their number density must lie within certain limits. If we were on other planets with earthlike molecular atmospheres, we might see orange or red skies, not blue. This makes us realize that the blue sky we see is special not only to human visual perception but to conditions on this particular planet.

These considerations all assume an atmosphere composed purely of molecules, but careful observations of skylight indicate significant discrepancies with pure molecular scattering. As was already clear in 1908, Rayleigh's scattering law does not agree perfectly with observation (see figure 10.1). Without denying the fundamental role played by molecular scattering, the sky's color has appreciable contributions from particles larger than molecules. The possibilities are complex and subtle. The basic facts are seen in the changing sky.

On hazy days of high ambient humidity, the sky's blue is often less deep, even rather whitish, which ought long ago have put to rest the recurrent theory that water droplets cause the clear sky's blue. Here, the crucial mistake was confusing a *gas*, water vapor (a small fraction of the molecular atmosphere), with *liquid* water droplets, whose varying presence and size have dramatic effects on the sky. A denser haze merges into fog, in which airborne water droplets can condense on surfaces so that we recognize that we are in a low-lying cloud. Up to now, we have excluded the influence of water droplets, seeking to determine the cause of the sky's blue even when ambient humidity is low. But clouds bear witness to the presence of water in the sky, to greater or lesser degrees.

Though it was clear from the earliest times that clouds are made of water, it took a long time to clarify the mechanisms responsible for their color. Aristotle already argued that the water droplets in the clouds act like tiny mirrors.[16] To this, many later accounts based on the wave theory added that, since these droplets are much bigger than the wavelength of light, they reflect all wavelengths equally and thus produce white light. In contrast, molecules, being much smaller than the wavelength of light, scatter shorter wavelengths preferentially.

Unfortunately, this elegant and persuasive explanation, though repeated in many texts, is misleading, as was emphasized by Craig Bohren in 1983.[17] It is true that cloud droplets are much bigger than the wavelength of light and do reflect light independently of its wavelength. But these are *sufficient* conditions for the whiteness of clouds and not *necessary* conditions.

To see this, go back to the experiment of adding a little skim milk to water to give a faint blue tint (experiment 5.6). That blue shows that the milk globules scatter smaller wavelengths more than they scatter longer ones. Hence, the globules are smaller than the wavelength of light and do not act as mirrors, as the water droplets in clouds are said to do. But now keep adding milk to the water, which will eventually turn *white*. The globules remain the same in size and composition. All that is different is their density, and that is crucial. In deriving his scattering law, Rayleigh had assumed that each wave undergoes only one scattering.[18] In the more dense milk solution, the light undergoes *multiple scattering*, which tends to equalize the scattering of all wavelengths. The reason is that, although a single scattering favors shorter wavelengths, this effect averages out when the light undergoes more and more scatterings. Compare two incident light waves, one of a bluer wavelength, the other redder. Though the bluer waves are more scattered than the redder on first scattering, as they both undergo more and more independent scatterings, the probability of scattering any given wavelength continues to add and grow. The result is an equal mixture: white.

Thus, multiple scattering *necessarily* produces white light and can cause dense clouds of small droplets to appear white. Multiple scattering also explains why the sky seems whiter at the horizon than at the zenith even on a clear, dry day: The light reaching us from the horizon passes obliquely through the atmosphere and hence undergoes many more scatterings than light coming from the zenith. This refutes the common belief that horizon whitening is due only to particles in the air, such as pollution. Though such particles can have that effect, they are not necessary. For instance, through extremely clear air in Santa Fe, I still see whitening on the horizon even when very little pollution is present.

To be sure, there are always particles in the air and yet the air is often blue. Yet particles can affect the color of the sky, depending on their composition.[19] In chapter 7, we noted that Tyndall's sky in a bottle produced only a yellow-orange "sunset" along the direction of the incident beam of light. A purely molecular atmosphere is insufficient to account for the deep red hues that make some sunsets so spectacular. Such sunsets can come from the presence of particles in the air caused by pollution, forest fires, dust storms, or volcanic explosions. These particles are generally much bigger than the wavelength of light and are collectively called aerosols. Their different sizes and compositions will give rise to different spectral patterns of scattering. For simplicity, we have treated all species of molecules as scattering light according to Rayleigh's universal law.

Though Rayleigh's law is correct for pure molecular scattering, the story does not end here. In 1953, E. O. Hulburt argued that the sky's zenith color at twilight depends strongly on the presence of a specific molecule: ozone.[20] Ozone is formed naturally in the upper atmosphere as a result of the interaction between the energetic ultraviolet radiation from the sun breaking the bonds of the normal molecular oxygen, O_2, which can then reform as ozone, O_3. This process does not normally happen in the lower atmosphere, because ozone absorbs most ultraviolet light before it reaches the ground. Ozone also absorbs light throughout the visible range, peaking in the red, near 600 nm. In contrast, neither normal molecular oxygen (O_2) nor nitrogen show such absorption of visible wavelengths. The ozone layer is high in the atmosphere and occupies only about 3 percent of the atmosphere's vertical depth; thus, it is not significant when the sun is overhead. But at sunset the light enters at a grazing angle, traversing a greater distance in the bands of ozone, which therefore are proportionally much more important than at midday.

Thus, just when the problem of the sky's color seemed settled, it reopened. In the 1950s, meteorologists reexamined the theory of the sunset, using the calculational power of the new electronic computers needed to deal with the complexity of the full theory. An inclusive understanding must involve not only gases of different molecules (especially ozone) but also aerosols. This required reaching beyond classical physics, which does not give an account of the behavior of molecules. Understanding the light absorption pattern of ozone (or of copper compounds) requires quantum theory, which describes how molecules absorb light of different wavelengths.

For instance, in 1974 Charles N. Adams, Gilbert N. Plass, and George W. Kattawar calculated the scattering of light in five different model atmospheres, each having different amounts of ozone and aerosols.[21] Their treatment was relatively simple, in that they assumed that each photon is scattered only once. Even so, their calculations required a substantial amount of time on the best computers of the period. In the model without ozone, they computed that the twilight zenith should look yellow or essentially white. When ozone was added in the normal amount, the zenith regained its blue. When the aerosols were three times normal (as tabulated from atmospheric data), the zenith became more intensely blue and the red and yellow area on the horizon became more intense and larger. When the aerosols were ten times normal, the zenith became a washed-out gray-blue and the sunset even more intensely red and yellow, expanding to an even larger area in the sky.

Thus, the zenith color of the sky at twilight depends on ozone. The depletion of the ozone layer is one of the most alarming and important problems affecting the earth.[22] Without the ozone layer, ultraviolet radiation at ground level would increase, bringing with it increased risk of cell destruction and skin cancer. During the late 1960s, it was realized that nuclear weapons testing and flight at very high altitudes could deplete the ozone layer, which gave additional impetus to the movement to cease those activities. In the 1970s, chlorofluorocarbons (CFCs) used in aerosols, refrigerants, and foam propellants were implicated as sources of chlorine that would tend to break down ozone. By the 1980s, there was evidence of an "ozone hole" above the South Pole, a region of depleted ozone and elevated chlorine concentrations.

Since then, the ozone hole has grown and there are signs of ozone depletion over the entire globe. Studies in Canada indicate greater amounts of ultraviolet light at the earth's surface. The oceans near Antarctica show depressed plankton photosynthesis, which also implies greater ultraviolet radiation there. Thus, the depletion of the ozone layer has profound effects on plant life and the ocean, with all that implies for animals and humans. The total depletion of ozone would probably make the earth much more inclement, if not totally uninhabitable for humankind.

These ominous effects of ozone depletion are of a different order of significance compared to its role in regulating the zenith color of the sky at twilight. Yet that color is the direct correlate of those same fundamental processes and is moreover directly before our eyes. Even more immediate is the danger posed by the aerosols poured into the air by automobiles and industry. Will it move people to think that the blue sky above them can be destroyed by human folly?

Ruskin's visions of a noxious "storm-cloud" anticipated our present peril. But any real solution will require the knowledge gained by studying the sky in a bottle, so that we can more knowingly limit human greed and overreaching. We now can gaze at the sky of Mars via remote television cameras on its surface. Lacking the earth's ozone layer and often dust-filled, the Martian sky is a dull ochre, through which the setting sun shines blue, as it did on earth after Krakatau (Krakatoa) exploded.[23] Perhaps it will give us pause to think that our children may inherit such a sky.

Beyond the conventions of romantic imagery, sky blue is imprinted in the human imagination. For Henry James, "the country of the blue" is the visionary realm, the true source of art.[24] Isak Dinesen tells of the Lady Helena, who searched the world for rare blue china, buying hundreds of blue jars and bowls only to put them aside, saying: "Alas, alas, it is not the right blue." When her father suggests to her that perhaps the color she seeks does not exist, she replies: "O God, Papa, how can you speak so wickedly? Surely there must be some of it left from the time when all the world was blue." At the end of her life, she finally finds the true blue in a jar: "Oh, how light it makes one. Oh, it is fresh as a breeze, as deep as a deep secret, as full as I say not what." If her heart is put into that perfect blue jar, "all shall be blue round me, and in the midst of the blue world my heart will be innocent and free."[25]

Has, then, the sky been captured in a bottle, as it was in Lady Helena's jar? Have we insulted Athena and violated her mysteries, as Ruskin suggested, so that we should now make amends and release her from captivity? I leave you to judge. For myself, I find it strange and beautiful that such simple questions lead to such deep realizations about the nature of the universe. These insights may lead us to look at the sky with new eyes and new questions.

The quest to understand the sky and its color leads inward, for the sky cannot be blue if atoms are not real. Gazing at the sky, we confront the most beautiful proof of atomic theory. The quest also leads outward, to the furthest galaxies, whose distribution determines the brightness of the night. Contemplating its light, we receive silent evidence of the universe's vastness in space and time. Between day and night is twilight. There, the zenith color tells of the fragile condition of the earth itself, poised between macrocosm and microcosm.

Appendix A: Experiments

Experiment 2.1 Ristoro and Leonardo's Recipes for Blue

To try Ristoro's technique, apply white oil paint to a small area of canvas. After it dries, take a small amount of transparent black paint on your brush and apply it *very* delicately over the white area, aiming for an extremely thin veil of overpainting. The result should be a delicate blue-gray, though my own attempts yielded only various shades of gray.

Leonardo says that painting a veil of transparent white over intense black yields a "beautiful blue," which is far more problematic. One modern scholar who tried this got only "unpleasant grays with a greenish cast," nor was I able to get anything but gray. Did Leonardo really carry out this experiment, or was he imagining the result on the basis of black over white? Surely pigment composition and artistic skill matter; a painter whom I asked was able to make white over black yield blue, but not vice versa. (See also p. 206, note 22.)

Experiment 3.1 Newton's "Blue of the First Order"

The easiest way to observe this phenomenon is to use a simple metal or plastic hoop, such as children dip into soap solutions to form bubbles. (The commercial bubble solutions usually contain a few drops of glycerin in order to stabilize the bubbles by increasing their surface tension.) Dip the hoop into the soap solution and watch the thin film. You will notice changing colors in the film, as Newton describes. Gradually, a dark blue (his "blue of the first order") will form under the top edge of the hoop, as gravity draws the soap solution away from it and makes that upper edge of the bubble ever thinner.

Experiment 3.2 Grimaldi's Diffraction

Grimaldi had to find pointlike sources of light and work in total darkness, so delicate are the phenomena of diffraction. We can take advantage of modern light sources, like the tiny incandescent light bulb in an ordinary high intensity flashlight or, better yet, an inexpensive laser pointer.

Take a pin (width about 0.5 mm) and illuminate it with the laser or flashlight. Darken the room and look at the shadow cast by the pin on a white, smooth surface about a foot (0.3 m) in front of the pin. On the edge of the pin's central shadow, you should see faint fringes of light, receding in brightness away from the center of the shadow (see Grimaldi's drawings in figure 3.6a). See also experiment 5.3 for an alternative way to see diffraction from a single slit.

Experiment 4.1 Saussure's Sky in a Bottle

The simplest way to produce such a solution is to pour some ordinary blue food coloring into water until you obtain a blue that seems to you comparable to sky blue. If you illuminate the solution with a beam of light, notice that the color looks the same viewed at any angle to the beam; this is because the color comes from *absorption* (the solution absorbs red wavelengths so that only blue remain). You may want to compare the appearance of the solution against different backgrounds, black or white. You can also add milk or chalk particles or even just bubbles (by stirring); notice how they increase the luminance (brightness) but decrease the purity of the color.

Duplicating Saussure's own experiment requires locating two common chemicals. Begin by finding some copper sulfate ($CuSO_4$), which can be obtained in crystalline form in children's chemistry sets or from chemical supply stores. It is also available in hardware stores as a compound sold to kill tree roots (for instance, in sewer pipes). Make a solution of copper sulfate by adding it to a small quantity of water; the solution is saturated when no further copper sulfate can be dissolved without precipitating out of solution. For $\frac{1}{2}$ cup (100 ml) of water, you will need about 1 ounce (32 gm) of copper sulfate crystals to make a saturated solution. Note its characteristic color.

The other chemical you will need is ordinary liquid ammonia (without any additives or tint), sold in most drugstores (the concentration ordinarily sold there will work well). Add the ammonia slowly to the saturated copper sulfate solution. At first you will see some precipitate formed, which dissolves as more ammonia is added. (*Important safety note*: Both copper sulfate and ammonia are toxic substances that must not be ingested.

Ammonia also has noxious fumes that should not be breathed for long. Accordingly, this experiment should be done outside or in a well-ventilated place, and children should not do it without close adult supervision.)

Notice the very deep blue of the suspension formed, in comparison to the more greenish color of the copper sulfate solution alone. If you wish, you can also try Saussure's suggestion of obtaining different shades of this blue by adding pure water. Also, try observing the color of the solution in strong light as opposed to very dim light. Compare this with the way peacock feathers look in strong versus dim light. The difference is color by *absorption* versus by *scattering* of light.

Experiment 5.1 Is Sky Blue an Optical Illusion?

Take a cardboard mailing tube and look with one eye through it at the blue sky, keeping your other eye open also. After a few seconds, the eye looking through the tube will see an increasingly whitened color, while the other eye continues to see the same blue. Georg Wilhelm Muncke (1820) took this as proof that the blue of the sky was an optical illusion, perhaps caused by the eyes perceiving the complementary color to sunlight. But now do the same experiment looking through the tube at some bright blue object, like a piece of blue cloth, which shows the same "bleaching" of the retinal receptors when they stare at a uniformly blue field of view. From this, Heinrich Wilhelm Brandes argued that sky blue is no less real than the blue of the cloth.

Experiment 5.2 Goethe's Urphenomenon

Make a dilute soap and water solution in a clear glass. Use a *clear, untinted* soap, such as liquid detergent; usually only a few drops are needed, though depending on its concentration, you may need to add more. Illuminate the glass from the front and view it against a dark background and you see a faint bluish-violet; against a brightly illuminated background, reddish-gold. Goethe also observed it by looking through the lower section of a candle flame. When seen against a dark background, the familiar blue color appears; against a bright background, the blue disappears.

Experiment 5.3 Young's Interference

Young had to pierce pinholes in his window blinds in order to create a pointlike source of an intense beam of light. If you don't want to do that, an ordinary incandescent high-intensity light will do, or a high-intensity flashlight; best of all is a laser pointer. (Do not look directly into laser or halogen lights.) To produce slits, take a microscope slide (needed just for mechanical support) and attach to it a piece of aluminum foil by using clear tape on the edges of the foil. Then cut a single slit into the foil using a razor blade or sharp knife. Project laser light through the slit onto a white background and notice that you see not just a central sharp slit but also a pattern of fringes around it; this is a nice way to see Grimaldi's diffraction (see experiment 3.2).

To make a double slit, cut a second slit as close as possible and parallel to your first slit (try for a spacing of 0.5 mm or less). If you make one of these slits about 0.5 cm longer than the other, you can easily shift between projecting the one-slit and two-slit patterns. Try also nonlaser lights of different colors, made by filters. Note that when you try these experiments with a broad source, such as the sky or an ordinary frosted light bulb, you won't see the interference patterns. (Why? We discuss this in chapter 8.) The subtlety of these phenomena explains why they were not clearly seen until about 1800.

Experiment 5.4 Seeing the Wave Nature of Light

There are simple ways to observe the wave nature of light without instruments. If you are in bright sunlight, half-close your eyes so that you are looking through your eyelashes. You will notice color fringes at the edges of your eyelashes. These are the direct result of the different wavelengths contained in sunlight diffracting around the eyelashes (which here are acting like Young's slits) and forming bright maxima at different places.

A similar effect occurs when you look at someone's hair brightly backlit, or at a field of thin, dry straw or grass lit by the sun low in the sky. You will see a glow surrounding each hair or straw. If light were particles, the hair or straw would show a simple, sharp edge. If light is a wave, it bends around those edges and interferes constructively to form the glow (as also in experiment 5.3).

Alternatively, take an ordinary CD and observe the colors produced by reflected white light. Since the surface of the CD is just a series of closely spaced reflecting grooves (a "diffraction grating"), then the different wavelengths within the white light are each interfering constructively at different angles, visible as the colored pattern.

Finally, the light from a laser pointer is highly coherent (its waves being closely correlated), so that shining it on an ordinary matte white surface gives clear indication of interference in the "grainy" stippling one sees near the central image of the beam. Had Newton seen such a light, he would have confronted dramatic evidence of its wave nature.

Experiment 5.5 Seeing Haidinger's Brush

Many people have trouble seeing this phenomenon simply looking at the sky with the naked eye (though a few see it directly). It helps to stare at a clear blue sky, near the zenith, through a polarizer (a sheet of the plastic polymer used to make polarized sunglasses) for five or ten seconds and then suddenly rotate the polarizer by 90°. The brush then appears more clearly. It will rotate if you turn the polarizer, and then it fades away rapidly. If you do not have access to sheets of polarizer, simply use polarized sunglasses.

Experiment 5.6 Seeing Blue in Suspensions

For an easy version of Brücke's experiment, take a clear glass with water and add a few drops of *nonfat* milk. This will give a distinctly blue tinge to the water; indeed, nonfat milk by itself has this blue hue, the result of small milk globules suspended in water. Alternatively, sprinkle a small amount of powdered nonfat milk or creamer onto the water and watch as it dissolves. The blue tinge will soon appear; I noticed lovely "snowstorm" effects of the flakes, which sometimes even seem to "fall" upward.

These are forms of Goethe's urphenomenon (experiment 5.2). Note that adding more milk turns the solution *white*, not blue, so that whole milk loses this blue tinge, compared to nonfat. As is discussed in chapter 10, the white color depends on *multiple scattering*: The light is scattering not just once but many times as it passes through the denser solution. Also try looking at the light from these suspensions through a polarizer, either Polaroid sheets (or sunglasses) or a calcite crystal.

Experiment 6.1 Tyndall's Sky in a Bottle

This experiment requires some special chemicals. Tyndall himself used butyl nitrite, benzene, and other highly toxic, carcinogenic substances, whose dangers were not appreciated at the time. However, an adaptation of one of Roscoe's experiments can be safely and readily duplicated. You will need to obtain sodium thiosulfate ($Na_2S_2O_3$), which used to be a common photographic fixative called "hypo." It has been superseded by other compounds, so it is not generally available in photographic supply stores but is easily obtained from chemical suppliers. You will also need a medium-sized (rectangular) fish tank about 1–2 feet long (0.3–0.6 m); mix 1 gallon (3.2 liters) of water with 20 gm of sodium thiosulfate and stir well to dissolve. You will need to arrange a bright (high-intensity) light to shine a small beam through this liquid; it is best if the beam is directed through a tube or baffle containing a convex lens (like a slide projector). The beam should shine through the liquid in the tank and emerge out the far (long) side of the tank onto a white surface; notice the initial color of this transmitted light. These observations are best done in a darkened room.

Now you will need to add 2 ml of concentrated sulfuric acid, readily available in a high school chemistry lab or from a chemical supplier. (Hydrochloric acid can be used instead.) Caution is required because of the acid's corrosive effects; such acids must always be added to water, not vice versa. About three minutes after adding the acid, you will notice a delicate precipitation and also a faint but noticeable blue tint of the beam, viewed from above or from the side. At the same time, the transmitted light will become yellow-orange, showing that the blue light is scattered from the beam. (Compare this to experiment 4.1, whose color is due to absorption, not scattering, and looks the same from any angle.) After about ten minutes, the precipitating particles will increase in size, so that the beam appears white, not blue; the transmitted light will then also become white again.

Appendix B: Letters on Sky Blue between George Gabriel Stokes, John Tyndall, and William Thomson[1]

[In a letter dated December 2, 1868 (RI MS JT/1/244), Tyndall tells Stokes the details of his experiments on light scattering through "artificial skies."]
December 7, 1868 [RI MS JT/1/S/246]

My dear Tyndall,

I think your paper certainly worthy of being communicated to the Royal Society. If the idea is new to so distinguished a physicist as Sir John Herschel, and has not been put in print by anyone, there is reason enough for producing it. The idea of accounting in this way for the blue of the sky was one which could hardly help forcing itself upon me, familiar as I was from my experiments in fluorescence with the beams produced by finely divided colourless precipitates, with the bluish appearance of such a mixture of reflected light, and with the polarization of the scattered beam.[2] But I never followed and [*sic*] the subject either theoretical (beyond giving the obvious explanation of the bluish rays) nor experimentally, and I have long regarded it as an undecided question whether this was the true account of the blue colour or whether there might not possibly be reflexion even within homogeneous air (i.e. air free from motes or haze) whether by reason of a minute reflexion even from the ultimate molecules, or from variations of density giving surfaces on opposite sides of which the medium had *slightly* different values of the refractive index. To disprove the existence of such reflexion in homogeneous (or moteless) air is a thing hardly within the reach of experiment considering how small are the thicknesses of air looked through in our experiments compared with the miles through which we look in regarding the sky.[3] Still if you can show by experiment that you can by a sufficiently fine precipitate get a blue *as deep* as that of an Italian sky, and that air can be so purged as to show no *sensible* beam that goes far to render improbable reflexion otherwise than by motes.

I have marked your letters with the dates of reception not knowing whether you mean the first of them for a communication to the Society or merely a sketch of a somewhat longer communication you mean to write on hearing from me.

Yours very truly
G. G. Stokes

December 5, 1868 [marked "rec[d] Dec 6/68"; RI MS JT/1/T/1400]

My dear Stokes,

Our apparent "enemies" are often our real friends. I thought for a time these particles of the atmosphere, and these clouds of decomposition, my worst enemies, and it was in my efforts to get rid of them that I changed my feelings towards them. . . .

[Tyndall then describes in detail his duplication of Govi's experiments passing a light beam through incense or gunpowder smoke and gives detailed data about the angle of polarization of scattered light.]

These are some of the effects which have caused me to change my feelings towards these motes & fumes. I entertain a strong opinion that as regards the blue of the sky the notion of reflection by the particles of the air themselves will prove to be an hypothesis that has not a single fact to support it; while abundant facts will be produced to show the sufficiency of a different theory. . . .

[Tyndall closes by proposing to communicate his observations to the Royal Society.]

December 11, 1868 [RI MS JT/1/S/248]

My dear Tyndall,

I thought I recollected talking to Roscoe about the cause of the blue of the sky, and stating my belief that it arose, in part at least, from suspended particles—whether over and above that there is such a thing as molecular reflection, of which *some portion* of the deeper blue may be due, I regard as a point which it is hardly in the power of experiment to settle. Yesterday I casually came across Roscoe's lecture at the Royal Institution (Proc. Vol IV),[4] on the opalescence of the atmosphere in which precisely that explanation is given. If I am right in supposing that you regarded this as the capital point of your communication, the discovery that it has already been given *in print* may modify the form in which you may think it right to bring forward your communication. As to the theory of the production of blue—why it is that finely suspended particles give a bluish reflexion, I think Brücke altogether wrong.[5]

I enclose a glass which it may interest you to examine. Please to let me have it again when you have done with it, but there is no hurry.

Yours very truly
G. G. Stokes

April 3, 1870 [RI MS JT/A/7C]

My dear Thomson.

I observed with immense pleasure the sweep of your scimitar through the atoms and molecules. To this we must come at last. There can be no permanent peace of mind until we know something definitely of the size and other physical characteristics of atoms and molecules, and to this you have made a contribution marked as might be expected by extraordinary penetration.[6] I have recently been engaged in colour experiments which have a general bearing upon this question and which I should therefore like highly to communicate to you.

One feature of the clouds produced by the decomposition of vapours by light is that you can regulate, at will, the growth of their particles, making it either slow or rapid by varying the quantity of matter acted on by the light. When the vapour is very attenuated the cloud first appears of a beautiful blue colour, and afterwards, in the great majority of cases, passes from blue through whitish blue to white. Did the particles first formed merely grow uniformly without the accession of new particles we should have a succession of colours, and it is (I suppose) because side by side with the growth of the larger particles new small ones are introduced, thus abolishing the predominance of any particular magnitudes, that the passage is, in most cases, that which I have indicated—namely from blue through whitish blue to white.

The depth of the colour is a rough measure of the relative size of the particles. If for example two nebulas be formed under precisely the same conditions with the sole difference that one has been acted upon for three minutes and the other for four minutes, then the relative grossness of the latter is marked by its whitish hue. You must not imagine that the particles are seen as floating motes. The light shed by the nebula is as continuous as the spectrum blue of incandescent carbon.

You know Brücke's receipt for precipitating mastic in water, so as to imitate the blue of the sky. Now when you find water containing such a precipitate whiter than an actinic cloud is it not fair to infer that its particles are coarser than those which constitute the cloud? I have prepared various precipitates which yield blue light and have placed them in the hands of eminent microscopists, but they can make nothing of the precipitated particles. The blue nebula is irresolvable by the highest microscopic powers. Huxley and Bence Jones have both sought for the particles, but the liquid in which they are suspended appears as free from particles as pure water.[7] The nebulous image appears perfectly homogenous.

Now let us return to the clouds. Imagine the attenuated vapour acted on by the light. For a time all shows darkness of pure transparency, but in two or three minutes a delicate blue appears at the most concentrated portion of the beam. At right angles to the beam the light discharged is perfectly polarized: the cloud can be quenched utterly by a Nicol's prism. Let the light continue to act. During its action the molecules fall in showers upon the particles. It is, on an atomic scale, your cannonade of meteors against the sun;[8] and yet, after twenty minutes or half an hour of such action the light scattered by the nebula in the direction of the normal remains perfectly polarized—the cloud can be utterly quenched by the Nicol. This, of itself indicates the amazing smallness of the particles. But we can also compare the colour of the nebula with that of the mastic precipitate, and infer that after the twenty minutes action the particles of the cloud are still smaller than the particles of the mastic. That is to say, smaller than particles which demonstrably defy the highest powers of the microscope. Now if this be the state of matter at the end of the 20 minutes, *what must be the size of the particles when they first appear?* And yet I suppose we must conclude that even then each particle is a heap of molecules and each molecule a heap of atoms. We may in time find out how many of the molecules thus grouped together are competent, under the conditions referred to, to send a sensible amount of light to the eye.

The escape of the normally scattered light from perfect polarization as the particles grow larger is beautifully marked. The short waves are those which may be expected to first *feel* the influence of size and accordingly, we find the first colour which reaches the eye when the Nicol is in its position of *minimum* transmission to be a perfectly gorgeous blue. Neither in the skies of the Alps nor in Italy have I seen any thing to compete with this "residual blue" in depth and luster. This also grows under your eyes to whiteness.

I have been deflected for a moment from my usual work to look into the question of "spontaneous generation." It is amusing to observe how much the notion of littleness has had to do with the diffusion of this doctrine. It was all but dead when the invention of the compound microscope revived it. The amazing smallness of the creatures revealed by this instrument gave the doctrine a new foothold. I say it is amusing because the observers seem not to have the least idea that in their vibrios[9] and bacteriums, they are dealing with perfect behemoths as compared that the *particles* of actinic clouds, not to say any thing about your world of *molecules* which lies behind those particles.

I was glad to hear from Mrs. King that you are well. The atoms indeed show that your mind, at all events, retains a firstrate edge and temper.

Believe me
Yours very sincerely
John Tyndall

How do you figure the content of your elastic molecular bags? Does not their power of restitution depend on the interaction of their parts? What are those parts?

Notes

References relevant to the whole of a chapter appear at the beginning of the corresponding section below, followed by specific endnotes. The abbreviation 2:345 means vol. 2, p. 345; in an ancient text, 3.456 means book 3, line 456. Please note that, in order to leave the text clear, at times citations are grouped at the end of the relevant paragraph of text.

After I wrote the first version of this book, I read Götz Hoeppe, *Blau: Die Farbe des Himmels* (Heidelberg: Spektrum Akademischer Verlag, 1999), an engaging treatment I recommend to German-speaking readers, especially for its fine color illustrations. My approach has been different, exploring some figures he did not include or discuss in detail (Kepler, Descartes, Leslie, Roscoe, Ruskin, Faraday, and Lodge), giving a larger account of Smoluchowski, treating in a new way the night sky, and at times offering different conclusions.

General works on atmospheric phenomena: Among historical treatments, Carl B. Boyer, *The Rainbow: From Myth to Mathematics* (Princeton: Princeton University Press, 1987) stands out for its unsurpassed scholarship and insight, as does Raymond L. Lee and Alistair B. Fraser, *The Rainbow Bridge: Rainbows in Art, Myth, and Science* (University Park: Pennsylvania State University Press, 2001), which is beautifully illustrated and particularly rich in its exploration of cross-cultural and artistic aspects. For another classic that presents a vast array of information about many atmospheric phenomena, see M. Minnaert, *The Nature of*

Light and Colour in the Open Air, tr. H. M. Kremer-Priest, rev. K. E. Brian Jay (New York: Dover, 1954). Also of great value as overviews are Robert Greenler, *Rainbows, Halos, and Glories* (Cambridge: Cambridge University Press, 1980), Aden and Marjorie Meinel, *Sunsets, Twilights, and Evening Skies* (Cambridge: Cambridge University Press, 1983), and David K. Lynch and William Livingston, *Color and Light in Nature* (Cambridge: Cambridge University Press, 1995), all superbly illustrated. A fine introduction to optics can be found in David S. Park, Dieter R. Brill, and David G. Stork, *Seeing the Light* (New York: Harper and Row, 1986), 347–354. For a survey of phenomena relating to color, see Hazel Rossotti, *Colour: Why the World Isn't Grey* (Princeton: Princeton University Press, 1983). For an authoritative text that gives excellent physical explanations along with the relevant mathematical details, see Craig F. Bohren and Eugene Clothiaux, *Fundamentals of Atmospheric Radiation* (New York: VCH-Wiley, 2005). I particularly recommend Craig Bohren's classic books, *Clouds in a Glass of Beer: Simple Experiments in Atmospheric Physics* (New York: Wiley, 1987) and *What Light through Yonder Window Breaks? More Experiments in Atmospheric Physics* (New York: Wiley, 1991), which are notable for their clarity, lucid exposition of the underlying physics, and debunking of many common misunderstandings.

Introduction

1. John Tyndall includes this quote from Herschel in his paper, "On the Blue Colour of the Sky, the Polarization of Skylight, and on the Polarization of Light by Cloudy Matter Generally," *Philosophical Magazine* 37, 384–394 (1869), at 388–389.

Chapter 1 Out of the Blue

General works on the history of theories of light: Vasco Ronchi, *The Nature of Light: An Historical Survey*, tr. V. Barocas (Cambridge: Harvard University Press, 1970), and an excellent work by David Park, *The Fire within the Eye* (Princeton: Princeton University Press, 1997).

1. For the poems mentioned using the phrase *cang tian* ("blue heaven"), see *The Book of Songs*, tr. Arthur Waley (New York: Grove Press, 1960),

Waley's poem numbers 273 (65), 151 (121), 278 (131), 282 (200), 302 (257), where the standard Mao numbers are given in parentheses. All these examples use this phrase in the context of lamentation or disaster. I thank Wan-go Weng for his kind help with these references and for the revised translation of 273 (65) cited in the text, which relies on the Han dynasty commentary of Zheng Xuan.

2. My account is based on what I myself saw in 2001 at Fengdu. In spite of the Yangzi bridge construction project, I understand that this temple has since been left intact, protected by a retaining wall.

3. See *The Origin of the Young God: Kālidāsa's* Kumārasaṃbhava, tr. Hank Heifetz (Berkeley: University of California Press, 1985), 42, 76, 95, 107, 118 (Śiva as Blue-Throated God), 88 (sky as blue as a sword). See also Irmtraud Schaarschmidt-Richter, "Blau in der ostasiatischen Kunst: Farbe der Konzentration—Farbe des Unbestimmten," in *Blau: Farbe der Ferne,* ed. Harns Gercke (Heidelberg: Verlag Das Wunderhorn, 1990), 57–70, which argues that in Japanese art blue is used to represent the concentration of the Buddha.

4. See Michel Pastoreau, *Blue: The History of a Color* (Princeton: Princeton University Press, 2001), 27, and Manilio Brusatin, *A History of Colors* (Boston: Shambhala, 1991), 27–30, 36–37 (ancient Britons).

5. See Hanna Erdmann, "Sinn und Gebrauch der Farbe Blau in der islamischen Welt," in *Blau: Farbe der Ferne,* 71–81, especially 79–80 (folklore about evil eye), 77 (Attar quote and blue as color of sadness).

6. Hesiod: *Theogony,* tr. Norman O. Brown (Indianapolis: Bobbs-Merrill, 1953), 56.

7. From a fragment from Aeschylus' lost play *The Danaids,* cited in *The Presocratics,* ed. Philip Wheelwright (New York: Odyssey Press, 1966), 27.

8. Ibid., 96, 99.

9. Ibid., 230 (Pythagorean verses), 152–153 (Empedocles).

10. See *Paulys Real-Encyclopedie,* 11:2238–2242 and Harold Osborne, "Colour Concepts of the Ancient Greeks," *British Journal of Aesthetics* 8, 269–283 (1968), who notes that "the Greeks were more concerned with

rich saturation of colour ('thrice dipped') than with hue" (275). On Gladstone's theory of Greek "color-blindness," see Pastoreau, *Blue*, 24–27. Brusatin, *History of Colors*, 28–29, interprets *kyanos* as a dark-green color, "the tone of the sea reflecting the sky at times when the water does not have that darker, more threatening Homeric color to it." However, the Liddell-Scott Greek lexicon glosses *kyanos* as dark blue, the color of lapis lazuli. (Note the alternative spelling, *kuanos*.) See also John Gage, *Color and Culture: Practice and Meaning from Antiquity to Abstraction* (Berkeley: University of California Press, 1993), 11–27.

11. See John Ruskin's 1869 lectures collected in *The Queen of the Air, Being a Study of the Greek Myths of Cloud and Storm*, available in *The Complete Works of John Ruskin*, Library Edition, ed. E. T. Cook and Alexander Wedderburn (London: George Allen, 1903–12), 39 vols., 19:279–423, specifically 306–307 (aegis and helmet), 328–329 (Athena coming through the window as air). I will return to Ruskin's ideas in chapter 6. Raymond E. Fitch, *The Poison Sky: Myth and Apocalypse in Ruskin* (Athens, Ohio: Ohio University Press, 1982), 532–575, gives a thorough discussion of Ruskin's treatment of Athena, as does Robert Hewison, *John Ruskin: The Argument of the Eye* (London: Thames and Hudson, 1976), 154–163, and Dinah Burch, *Ruskin's Myths* (Oxford: Clarendon Press, 1988), 93–131.

12. Homer: *Iliad* 11.23–46, using line numbers of the Greek text, cited from Homer, *The Iliad*, tr. Robert Fagles (New York: Viking, 1990), 297.

13. Ibid., 16.66 (the black cloud of Trojans); 20.418 (the dark mist of death); 11.629 (blue feet of table).

14. Homer: *Odyssey* 7.87, tr. Robert Fagles (New York: Viking, 1990), 182.

15. Plato: *Timaeus* 68c–d. In *Phaedo* 109e–110a, Socrates suggests that "we call the air heaven, as though it were the heaven through which the stars move" and imagines that if someone could take wing and rise to the top of the atmosphere, "he would recognize that this is the true heaven and the true earth." Though he does not explicitly draw this conclusion, Socrates' argument seems to imply that the true color of the sky is black, if seen above the atmosphere, so that the blue we see is a

kind of impurity. But Socrates' remarks do not contain any explicit consideration of the problem of the blue sky.

16. Plato, *Timaeus*, 45b–e. The term "extromission" is sometimes used for this "active" view of vision, as "intromission" is used for the "passive" alternative.

17. David C. Lindberg, *John Pecham and the Science of Optics:* Perspectiva Communis (Madison: University of Wisconsin Press, 1970), 129. For the significance of this theory of vision for Plato's larger philosophy, see my essay "Seeing the Forms" (unpublished).

18. For the distinction between *lux* and *lumen* (derived from the Greek *phōs* and *leukos*, parallel to Italian *luce* and *lume*), see Ronchi, *Nature of Light*, 32, 60–67, including a discussion of *Fiat lux* in the fifteenth-century encyclopedic *Margarita philosophica* on 67. However, there is no such distinction implied in the Hebrew word *'or* used in Genesis 1:3, as I learned thanks to Robert Sacks.

19. Empedocles: *The Presocratics*, 151 (on vision and effluences), 152–153 (on the sky). Plato discusses this view in *Meno* 76c, 75c–76e. See also Park, *The Fire within the Eye*, 39–50.

20. Lucretius, *On the Nature of Things*, 4.33–41.

21. Aristotle: *Sense and Sensibilia* 2.437b11–438b4, 3.440a15–19. All my quotations from Aristotle will use the translations in *The Complete Works of Aristotle*, ed. Jonathan Barnes (Princeton: Princeton University Press, 1984).

22. Aristotle, *On the Soul* 2.418b4–20.

23. Peter Pesic, "The Fields of Light," *St. John's Review* 38(3), 1–16 (1988–1989).

24. Aristotle, *Meteorology*, 371b19–375b15. For ancient meteorology, see H. Howard Frisinger, *The History of Meteorology: To 1800* (New York: Science History Publications, 1977), 1–23, Vladimir Janković, *Reading the Skies: A Cultural History of English Weather, 1650–1820* (Chicago: University of Chicago Press, 2000), 14–22 (on the changing connotations

of "meteors"), and Liba Taub, *Ancient Meteorology* (London: Routledge, 2003).

25. Aristotle, *On Colors* 1.791a2, 2.792a16–20. Pastoreau, *Blue*, 185, notes that this work is not by either Aristotle or Theophrastus.

26. Ibid., 3.793b14–794a15.

27. Aristotle, *Sense and Sensibilia* 3.439b20–440a15.

28. Wheelwright, *The Presocratics*, 178.

29. Lucretius, *On the Nature of Things*, 6.44, 6.96, 6.400.

Chapter 2 Ultramarine

1. See Gershom G. Scholem, *Major Trends in Jewish Mysticism* (New York: Schocken, 1965), 49–57, and William H. Donahue, *The Dissolution of the Celestial Spheres 1595–1650* (New York: Arno Press, 1981). In the sixteenth century, Averroes was cited as believing that the heavens were pure form without matter, as opposed to the prevalent view that the heavens were material. For Ptolemy's earlier view of the spheres as mathematical constructs, see *Ptolemy's Almagest*, tr. G. J. Toomer (Princeton: Princeton University Press, 1998), 35–37, 419–420; for his later view implying the physical reality of planetary spheres, see Bernard R. Goldstein, *The Arabic Version of Ptolemy's* Planetary Hypotheses (Philadelphia: American Philosophical Society, 1967), 7–8. I thank William Donahue for these references and helpful advice.

2. See Joseph Needham and Wang Ling, *Science and Civilisation in China* (Cambridge: Cambridge University Press, 1959), 3:462–494, at 479.

3. I have drawn from Howard R. Turner, *Science in Medieval Islam* (Austin: University of Texas Press, 1995), 17 (*ḥadīth* concerning knowledge), 20–21 (al-Kindī and his successors), 195–200 (optics). See also Seyyed Hossein Nasr, *An Introduction to Islamic Cosmological Doctrines* (Cambridge: Harvard University Press, 1964), 105–174 (Al-Bīrūnī), 174–274 (Ibn Sina), and *Religion, Learning, and Science in the 'Abbasid Period*, ed. M. J. L. Young, J. D. Latham, and R. B. Serjeant (Cambridge:

Cambridge University Press, 1990), 248–260, 274–289 (mathematics, physics, and astronomy), 364–369 (Al-Kindī), 405–423 (Al-Bīrūnī). On the use of colors in Islam, see Gage, *Color and Culture*, 61–64.

4. See Otto Spies, "Al-Kindi's Treatise on the Cause of the Blue Colour of the Sky," *Journal of the Bombay Branch of the Royal Asiatic Society* (new series) 13, 7–19 (1937), which includes the Arabic text and translation of al-Kindī's letter and also of the critical passage from Ibn al-Haytham's *Optics*. See also David C. Lindberg, *Theories of Vision from Al-Kindi to Kepler* (Chicago: University of Chicago Press, 1976), 18–32.

5. Ibn al-Haytham's great work is available in *The Optics of Ibn al-Haytham*, tr. A. I. Sabra (London: Warburg Institute, 1989), 1:13–38 (straight-line ray propagation experiments). For background and commentary see also Park, *The Fire within the Eye*, 78–80, H. J. J. Winter, "The Optical Researches of Ibn al-Haitham," *Centaurus* 3, 190–210 (1954), A. I. Sabra, *Optics, Astronomy, and Logic: Studies in Arabic Science and Philosophy* (Aldershot: Variorum, 1994), papers II–XI, especially paper VI on al-Haytham's concept of experiment. See also Ronchi, *Nature of Light*, 45–57; Matthias Schramm, *Ibn al-Haythams Weg zur Physik* (Wiesbaden: 1963); *Ibn al-Haitham: Proceedings of the Celebrations of the 1000th Anniversary*, ed. Hakim Mohammed Said (Pakistan: Hamdard National Foundation, 1969); and Saleh Beshara Omar, *Ibn al Haytham's Optics: A Study in the Origins of Experimental Science* (Minneapolis: Bibliotheca Islamica, 1977). See also Lindberg, *Theories of Vision*, 58–86. Regarding his work on twilight and atmospheric refraction, see A. I. Sabra, "The Authorship of the *Liber de crepusculis*, an Eleventh-Century Work on Atmospheric Refraction," *Isis* 58, 77–85 (1967).

6. Abu al-Rayḥān Muḥammad Ibn Aḥmad al-Bīrūnī, *The Exhaustive Treatise on Shadows*, tr. E. S. Kennedy (Aleppo: Institute for the History of Arabic Science), 31.

7. Hoeppe, *Blau*, 36–38.

8. Ronchi, *Nature of Light*, 57.

9. *The Opus Majus of Roger Bacon*, tr. Robert Belle Burke (New York: Russell & Russell, 1962), 2:481–482. Park, *The Fire within the Eye*,

110–111; Hoeppe has a somewhat different reaction to Bacon in *Blau*, 41–43. See also Lindberg, *Theories of Vision*, 104–116.

10. See Pastoreau, *Blue*, 32–83, and Brusatin, *History of Colors*, 43–50, and Beate Bender, "Color caelestis: Anmerkungen zur Farbe Blau im Mittelalter," in *Blau: Farbe der Ferne*, 82–103, especially 90–94 (Suger's light-theology), 99–101 (blue of demons and sinners). For Pseudo-Dionysius, see Dionysius the Areopagite, *The Divine Names and the Mystical Theology*, tr. C. E. Rolt (London: SPCK, 1940), 192–196, and John Gage, *Color and Meaning: Art, Science, and Symbolism* (Berkeley: University of California Press, 1999), 75–76. For the two kinds of theophany, see St. Gregory of Nyssa, *The Life of Moses* (New York: Paulist Press, 1978), 59–63, 94–97.

11. For God's identification with space and light, see Max Jammer, *Concepts of Space* (Cambridge: Harvard University Press, 1954), 34–47; note also the Jewish use of "place" (*makom*) as a name of God (26–31). The "dark angel" is found in the tenth mosaic of the north wall of San Apollinare Nuovo; see the illustration in Munemoto Yanagi et al., *Byzantium* (Secaucus; Chartwell Books, 1978), 32, and E. Kirschbaum, "L'Angelo rosso e l'angelo purchino," *Rivista di Archeologia Christiana* 17, 209–248 (1940), discussed along with other medieval uses of blue in Gage, *Color and Meaning*, 73–76. Colors of the divine light: Patrik Reuterswärd in *Light: from Aten to Laser*, ed. J. Hess and T. B. Ashbury; *Art News Annual* 35, 109ff (1969); and Gage, *Color and Culture*, 58–61.

12. Exodus 24.9–10 and Ezekiel 1.26. Ludovicus ab Alcasar: Bender, "Color caelestis," 97, and Hoeppe, *Blau*, 39. For a reflection on the significance of this sapphire blue in early Christian thought, see *Evagrius of Pontus: The Greek Ascetic Corpus*, tr. Robert E. Sinkewicz (Oxford: Oxford University Press, 2003), 180, 284.

13. Brusatin, *History of Colors*, 50–51, 58–59; Pastoreau, *Blue*, 21–24. See also Moshe Barasch, *Light and Color in the Italian Renaissance Theory of Art* (New York: New York University Press, 1978), 11–32 (Alberti), 44–89 (Leonardo), 135–209 (Lomazzo).

14. Ristoro (or Restoro) d'Arezzo: *La Composizione del Mondo*, ed. Alberto Morino (Parma: Fondazione Pietro Bembo, 1997), 4–5. The introduction

to this edition notices Ristoro's relation to Arabic Aristotelians such as al-Farghānī (xx–xxi). My account of Ristoro's work is indebted to Francesco Rodolico's article "Ristoro (or Restoro) d'Arezzo" in the *Dictionary of Scientific Biography* 11.468–469, whose translations I have adapted. I thank Franco Ligabue for his helpful advice with these translations.

15. Ristoro, *La Composizione del Mondo,* 294–295.

16. Ibid., 37.

17. Ibid., 350–352; see also Gage, *Color and Culture,* 133.

18. Leonardo da Vinci, *The Codex Leicester—Notebook of a Genius* (Sydney: Powerhouse Publishing, 2000), 50, (sheet 4A, folio 4r). The editor also notes Leonardo's debt to Ristoro. Even scholarly accounts of the history of light scattering tend to ignore this matter. For a recent popular account of Leonardo that ignores Ristoro, see Michael White, *Leonardo: The First Scientist* (New York: St. Martin's Press, 2000), 187–188: after mentioning Rayleigh's theory of light scattering leading to the understanding of the blue sky (1871), White asserts that "staggeringly, Leonardo got there three hundred years before him." I hope my treatment sets this canard straight. J. D. Hey also treats Leonardo as the starting point in his extremely helpful series of historical papers "From Leonardo to the Graser: Light Scattering in Historical Perspective," *South African Journal of Science,* Parts I and II, 79, 11–27, 310–324 (1983); Parts III and IV, 81, 77–91, 601–613 (1985). Though at one point Leonardo uses the word *attomi,* his subsequent comments show that his "minute, imperceptible particles" are not atoms but "particles of moisture" (*corpuscula dell'umidità*).

19. See the classic study by Marjorie Hope Nicolson, *Mountain Gloom and Mountain Glory: The Development of the Aesthetics of the Infinite* (Seattle: University of Washington Press, 1997), 4–6 (Ruskin), 28, 76 (Donne), 48 (Dante), 82 (blood of Abel). Donne's lines come from "The First Anniversarie. An Anatomy of the World," ll. 300–301. Nicolson connects the beginnings of the appreciation of mountains with the development of the new philosophy, specifically with Isaac Newton's teacher, the Cambridge Platonist Henry More (113–143).

20. See Hoeppe, *Blau,* 49–50.

21. Leonardo da Vinci, *Codex Leicester,* 158 (sheet 17B, folio 20r).

22. For the difficulty of duplicating Leonardo's claim about blue resulting from white over black, see C. Maltese, "Leonardo e la teoria dei colori," *Römisches Jahrbuch für Kunstgeschichte* 20 (1983). See the helpful discussion in Gage, *Color and Culture,* 133–137, especially for the contrast with Ristoro. For Titian's use of overpainting for blue, see Gage, *Color and Culture,* 134, and M. Jaffé and K. Groen, "Titian's 'Tarquin and Lucretia' in the Fitzwilliam," *Burlington's Magazine* 129 (1987).

23. *The Literary Works of Leonardo da Vinci,* ed. Jean Paul Richter (Berkeley: University of California Press, 1977), 1:161–163. See Brusatin, *History of Colors,* 66–67, 74–76. Though Hoeppe says that Leonardo considered air to be blue (as had Leon Batista Alberti about 1435), the passage quoted from the Codex Leicester ("I say that the blueness . . .", 50) indicates clearly that he thought it "not an intrinsic color." For Leonardo's references to Pecham, see Lindberg, *John Pecham,* 32. For Leonardo's theory of vision, see Lindberg, *Theories of Vision,* 154–168.

24. *Literary Works of Leonardo da Vinci,* 1:120, 226.

25. Alexander Moszkowski, *Conversations with Einstein* (London: Sidgwick & Jackson, 1972), 51–52.

26. Leonardo recorded his observations of light passing through water-filled vessels in his Manuscript F, f. 33v; see White, *Leonardo,* 187.

Chapter 3 Peacock Blue

1. See Edward Harrison's classic treatment, *Darkness at Night* (Cambridge: Harvard University Press, 1987), 34–37 (Digges's original statement of the puzzle), and 45–52 (Kepler's presentation), citing *Kepler's Conversation with Galileo's Sidereal Messenger,* tr. Edward Rosen (New York: Johnson Reprint, 1965), 34–36, 43.

2. *Kepler's Conversation with Galileo's Sidereal Messenger,* 18. For his treatment of the celestial spheres, see Donahue, *The Dissolution of the Celestial Spheres,* 93–100. See also Park, *The Fire within the Eye,* 160–164. For

an invaluable presentation of Kepler's optical views, see Johannes Kepler, *Optics: Paralipomena to Witelo & Optical Part of Astronomy*, tr. William H. Donahue (Santa Fe: Green Lion Press, 2000), 155 (on twilight), 282–287 (the redness of the eclipsed moon), though there is no specific discussion of the sky's blueness. For Kepler's theory of vision, see Lindberg, *Theories of Vision*, 178–208.

3. *Kepler's Conversation with Galileo's Sidereal Messenger*, 83 n. 131, the latter from Kepler's *Tertius Interveniens*, Thesis 49, reiterated in his *Epitome of Copernican Astronomy*.

4. In his *Discours de la nature de l'air* (1676), (Paris: Gauthier-Villars, 1923), Mariotte stated that "the elastic force of the air grows with the weight with which it is charged," often called "Boyle's law" in the English-speaking world. In that work, Mariotte also argued that air is really blue, not clear (61–62). His arguments are similar to those of Kepler, suggesting some familiarity with his optical writings. But Mariotte also offers an original argument: if you compare moonlight and candlelight shining on white paper, the moonlight seems distinctly blue. Mariotte does not attempt to connect what he considers intrinsic blueness with atomic theory; compare his conjecture that in the air are "an infinity of corpuscles so made that two or three coming together can initiate a plant and give it seed" (98). Since Mariotte considered air simply blue, I cannot concur with W. E. Knowles Middleton, "Random Reflections on the History of Atmospheric Optics," *Journal of the Optical Society of America* 50, 97–100 (1960), who includes him among those who held "that the blue of the sky was due to the clear air itself," naming also Bouguer, Euler, Leslie, and Saussure. I am grateful to Middleton for leading me to investigate these figures further. His paper is also available in an excellent collection, *Scattering in the Atmosphere*, ed. Craig F. Bohren (Bellingham, Wash.: SPIE Optical Engineering Press, 1989), 3–6.

5. See Janis C. Bell, "Zaccolini's Theory of Color Perspective," *Art Bulletin* 75, 91–112 (1993), which discusses his relation to Kepler in n. 16 and emphasizes his influence on Nicolas Poussin, whose cobalt-hued skies are notable; to what extent might this reflect Poussin's adherence to Zaccolini's views?

6. Sir Arthur Conan Doyle, "The Sign of Four," in *The Annotated Sherlock Holmes*, ed. William S. Baring-Gould (New York: Clarkson N. Potter, 1967), 1:638.

7. *Discourse on Method, Optics, Geometry, and Meteorology*, tr. Paul J. Olscamp (Indianapolis: Bobbs-Merrill, 1965), 344–345. For a superb account that places the *Meteorology* in the full context of Descartes's life and works, see Stephen Gaukroger, *Descartes: An Intellectual Biography* (Oxford: Clarendon Press, 1995), 58, 218–220, 295–296. Note that most modern editions of the *Discourse on Method* do not include the *Optics, Geometry*, or *Meterology*, though Descartes clearly meant these three specific works to accompany it. For Descartes's optical theory, see Boyer, *Rainbow*, 200–219; Ronchi, *The Nature of Light*, 112–119; and the especially ample treatment in A. I. Sabra, *Theories of Light from Descartes to Newton* (Cambridge: Cambridge University Press, 1981), 17–135. For Descartes in the context of earlier speculations on meterology, see Frisinger, *The History of Meteorology*, 24–43.

8. See Bacon's *New Atlantis* (1624), in *The Works of Francis Bacon*, ed. James Spedding (London: Longmans, 1857–1874 [reprint: New York: Garrett Press, 1968]), 3:127–166 at 137; David Renaker, "A Miracle of Engineering: The Conversion of Bensalem in Francis Bacon's *New Atlantis*," *Studies in Philology* 87, 181–193 (1990); and my book *Labyrinth: A Search for the Hidden Meaning of Science* (Cambridge, Mass: MIT Press, 2000), 47–55.

9. Descartes, *Discourse*, 69–70.

10. Ibid., 67–68.

11. Ibid., 71–72.

12. Ibid., 346–348.

13. For his critique of Descartes, see Isaac Newton, *The* Principia: *Mathematical Principles of Natural Philosophy*, tr. I. Bernard Cohen and Anne Whitman (Berkeley: University of California Press, 1999), 393–397, 779–790. For his violent aversion to Descartes's name, see Richard S. Westfall, *Never at Rest: A Biography of Isaac Newton* (Cambridge: Cambridge University Press, 1987), 401. On Newton's optical theories,

see Ronchi, *Nature of Light*, 160–208; Sabra, *Theories of Light*, 231–342; and Boyer, *Rainbow*, 233–257. For helpful commentaries, see A. Rupert Hall, *All Was Light: An Introduction to Newton's* Opticks (Oxford: Clarendon Press, 1993), Dennis L. Sepper, *Newton's Optical Writings: A Guided Study* (New Brunswick: Rutgers University Press, 1994), and Alan E. Shapiro, *Fits, Passions, and Paroxysms: Physics, Method, and Chemistry and Newton's Theories of Colored Bodies and Fits of Easy Reflection* (Cambridge: Cambridge University Press, 1993). For the earlier printed versions of his optical writings, see *Isaac Newton's Papers & Letters on Natural Philosophy*, ed. I. Bernard Cohen (Cambridge, Mass.: Harvard University Press, 1958), 47–59 (white light contains all colors), 194–199, 202–235 (Newton's rings), 232 (sky blue).

14. Sir Isaac Newton, *Opticks*, fourth ed. (New York: Dover, 1952), 1.

15. Ibid., 168–173, and Boyer, *Rainbow*, 251–257.

16. Robert Boyle, *Experiments and Considerations Touching Colours* (London, 1664; reprint, New York: Johnson Reprint, 1964), 244–245.

17. Robert Hooke, *Micrographia* (London, 1665; reprint, New York: Dover, 1961), 47–67.

18. Newton, *Opticks*, 210.

19. Ibid., 212, 372–373.

20. Ibid., 214–216.

21. Ibid., 216, 257.

22. Voltaire, *The Elements of Sir Isaac Newton's Philosophy*, tr. John Hanna (London: Frank Cass and Company, 1967 [reprint of 1738 edition]), 164. Voltaire devotes much of this popular account to Newton's optics.

23. Newton, *Opticks*, 345–346; for the "fits," see Shapiro, *Fits, Passions, and Paroxysms*, 136–207.

24. For excerpts from the experiments discussed here, see William Francis Magie, *A Source Book in Physics* (New York: McGraw-Hill, 1935), 294–298. There is no English translation of Grimaldi's whole book, but a reprint is available as Franciso Maria Grimaldi, *Physico-Mathesis de*

Lumine Coloribus et Iride (1665) (London: Dawsons of Pall Mall, 1966), 1–11.

25. Hooke, *Micrographia*, 57.

26. *The Correspondence of Isaac Newton*, ed. H. W. Turnbull (Cambridge: Cambridge University Press, 1959), 1:198–205, Hooke's letter to Lord Brouncker (June 1672). Though there was bad blood between them and Hooke was a hunchback, I do not think that Newton meant his reference to "the shoulders of Giants" to be an oblique, cruel gibe at his rival's stature, as some have interpreted. Newton observed the civilities of his day, whatever his inner feelings; when he attacked, he did so directly, not through low insinuations. See Newton, *Correspondence*, 1:416, letter to Hooke of February 5, 1676, and Westfall, *Never at Rest*, 272–274 and n. 106, who does not accept the sarcastic interpretation. Indeed, "standing on the shoulders of giants" was a commonplace with a long history, going back to Bernard of Chartres (†1126). See Robert K. Merton, *On the Shoulders of Giants: A Shandean Postscript* (Chicago: University of Chicago Press, 1993).

27. For Newton's experiments following "Grimaldo," see the *Opticks*, 317–339, Ronchi, *Nature of Light*, 124–149, Park, *The Fire within the Eye*, 189–193, and Alan E. Shapiro, "Newton's Experiments on Diffraction and the Delayed Publication of the *Opticks*," in *Isaac Newton's Natural Philosophy*, ed. Jed Z. Buchwald and I. Bernard Cohen (Cambridge, Mass.: MIT Press, 2001), 47–76.

Chapter 4 Shades of Blue

1. For Funck, Noellet, Hassenfratz, see Hoeppe, *Blau*, 80–84.

2. See *Pierre Bouguer's Optical Treatise on the Gradation of Light*, tr. W. E. Knowles Middleton (Toronto: University of Toronto Press, 1961), v–xii (general survey of his life), 232–235 (extinction), 234 (reflection by molecules), 240 (real separation of red and blue).

3. See the excellent edition of Johann Heinrich Lambert, *Photometry, or on the Measure and Gradations of Light, Colors, and Shade*, ed. and tr. David

L. DiLaura (New York: Illuminating Engineering Society of North America, 2001), 294 (transparent bodies swarm with hetereogenous particles), 308 (sea and sky); this edition also contains a very helpful survey of the history of photometry (xxvii–lxxxiv). In "Sur la perspective aërienne," *Noveaux Mémoires de l'Académie Royale des Sciences et Belles-Lettres* (Berlin), 74–77 (1774), Lambert asserts that sky blue comes from "solar rays reflected in the air," which is "filled with extraneous particles" (*fort chargé de particules étrangeres*), as if even clear air were "like a very fine and delicate fog" (*comme un brouillard très mince & très délié*). There is no evidence that Lambert was an atomist or acquainted with Roger Boscovich's notion of mathematical centers of force that could act as atoms. I thank David DiLaura for helpful discussion of Lambert's views.

4. Cited in Carl B. Boyer, *A History of Mathematics*, rev. Uta C. Merzbach, second edition (New York: Wiley, 1991), 440.

5. For Euler's extensive contributions to meteorology, see Frisinger, *The History of Meteorology*, 138–140; for his atomic theory, see Park, *The Fire within the Eye*, 235–237. For the letters cited in the text, see Leonhard Euler, *Letters of Euler on Different Subjects in Natural Philosophy, Addressed to a German Princess*, ed. David Brewster (New York: Harper, 1837), 1:125–128.

6. *Letters of Euler*, 2:407–410.

7. Ibid., 2:35.

8. I have relied on the entry by Albert V. Carozzi on "Horace Bénédict de Saussure" in the *Dictionary of Scientific Biography*, 12:119–123, as well as on the still useful biography by Douglas W. Freshfield, *The Life of Horace Benedict de Saussure* (London: Edward Arnold, 1920), 8–11 (Gesner quote).

9. Freshfield, *Life of Saussure*, 35–51 (attitudes toward the *Monts Maudits*), 23, 68 (Saussure as author of Alpine exploration and passion for Alpine scenery), 75 (work on rainbows), 87 (experimental program), 208 (observations with his instruments), 434–435 (invention of hygrometer and other meteorological instruments, including cyanometer and diaphanometer), 220, 258 (contrast of snow with blue sky), 149

(clarity of skies). Note also the useful "Note on the Meteorological Work and Observations on Deep Temperatures of H. B. de Saussure" by H. R. Mill, included as 457–465 in Freshfield, *Life of Saussure,* and Frisinger, *The History of Meteorology,* 85–89.

10. Saussure, *Voyages,* 2:550, translated in John Pinkerton, *A General Collection of the Best and Most Interesting Voyages and Travels in All Parts of the World* (Philadelphia: Kimber and Conrad, 1811), 4:798.

11. Freshfield, *Life of Saussure,* 80, 202. For Kant on the sublime, see Immanuel Kant, *Critique of Judgement,* tr. J. H. Bernard (New York: Collier Macmillan, 1951), 82–106.

12. Freshfield, *Life of Saussure,* 288.

13. Horace Bénédict de Saussure, "Description d'un cyanomètre ou d'un appareil destiné à mesurer l'intensité de la couleur bleue du ciel" and "Description d'un diaphanomètre ou d'un appareil destiné à mesurer la transparence de l'air," *Memorie della R. Accademia delle scienze di Torino* 4, 409–424, 425–453 (1788–1789). Saussure's diaphanometer works on a similar principle: He used a black circle on a white background, moved further and further away until he could observe the distance at which they are no longer distinguishable. He also noted even better results with approaching the circle from a great distance and noticing at what distance it becomes just distinguishable from its white background. These distances give a measure of the transparency of the ambient air. Saussure also constructed a series of circles of different diameters in a graded set to make comparative observations. To make a cyanometer for children, see the simple directions in Glen Vecchione, *100 First-Prize Make-It-Yourself Science Fair Projects* (New York: Sterling Publishing, 1998), 14–17, including suggestions for using it in observations of the sky.

14. Bouguer, *Optical Treatise,* 239: "It would be suitable to use prismatic colors as a reference, or at least to use boards painted in different colors," though he gives no evidence of having done so himself.

15. Saussure, "Description d'un cyanomètre," 410–411.

16. Ibid., 413–414, 417–424.

17. Ibid., 414–415.

18. Ibid., 416–417.

19. See David V. Thompson, Jr., *The Materials and Techniques of Medieval Painting* (New York: Dover, 1956), 153–154: "there is good reason to believe that most of these copper blues [used by medieval artisans], if not indeed all of them, were really copper-ammonia blue." I thank Lawrence Principe for drawing this reference to my attention.

20. Middleton incorrectly states that Saussure was of "the very sound opinion that the air molecules scatter more blue light than yellow or red," in "Random Reflections," 97; apparently, Middleton interpreted "elements" as "molecules," whereas Saussure's "Description d'un cyanomètre," 414, specifies them as "opaque and colored vapors."

21. In 1800, John Herschel wrote that "The rays of heat are . . . less refrangible than those of light; and . . . they are also, if I may introduce a convenient term, less scatterable," *Philosophical Transations of the Royal Society (London)*, 90, 523 (1800). See "scatterable, *a.*" *Oxford English Dictionary*, second ed., 14:599. In 1808, Herschel also used the word "scattering" in the context of optics (14:600).

22. For Saussure's relation to Humboldt and Darwin, see Freshfield, *Life of Saussure*, 440. On Alexander von Humboldt, see Douglas Botting, *Humboldt and the Cosmos* (New York: Harper and Row, 1973); L. Kellner, *Alexander von Humboldt* (London: Oxford University Press, 1963); and Helmut de Terra, *Humboldt: The Life and Times of Alexander von Humboldt 1769–1859* (New York: Knopf, 1955). For Humboldt's instrument list, see Hanno Beck, *Alexander von Humboldt* (Wiesbaden: Steiner Verlag, 1959–1962), 2:280–281. See also Alexander von Humboldt, *A Personal Narrative of a Journey to the Equinoctial Regions of the New Continent*, ed. Jason Wilson (London: Penguin, 1995), ix, xliii, xlv.

23. See Aaron J. Ihde, *The Deveopment of Modern Chemistry* (New York: Harper and Row, 1964), 46–47; for Priestley's original description, see *A*

Scientific Autobiography of Joseph Priestley (1733–1804), ed. Robert E. Schofield (Cambridge, Mass.: MIT Press, 1966), 161–162. See also Jean Joseph Gay-Lussac, "Expériences sur les moyen eudiométriques et sur la proportion des principes constituants de l'atmosphère," *Journal de physique* 60, 129–168 (1804) and M. P. Crosland, "Jean Joseph Gay-Lussac," *Dictionary of Scientific Biography* 5:317–327, at 318.

24. Alexander von Humboldt, *Cosmos* (Baltimore: Johns Hopkins University Press, 1997), 317–319.

25. Ibid., 323–324 (elements of climatology), 330 (isothermal lines), 306–307 (aerial ocean).

26. Alexander von Humboldt, *Reise auf dem Río Magdalena, durch die Anden und Mexico* (Berlin: Akademie-Verlag, 1986), 1:107, 147, 174, 181, 184.

27. Botting, *Humboldt*, 161.

28. Ibid., 95 (two suns); Kellner, *Humboldt*, 39–40 (straw-colored light; deep blue becomes lighter after rain).

29. Humboldt, *Cosmos*, 126–134. On zodiacal light, see also Meinel, *Sunsets, Twilights, and Evening Skies*, 91–100, and F. E. Roach and Janet L. Gordon, *The Light of the Night Sky* (Dordrecht: D. Reidel, 1973), 37–81.

30. *Cosmos*, 140–143; for his relation to the dark night sky puzzle, see Harrison, *Darkness at Night*, 144, 148 (regarding Edgar Allen Poe's admiration for Humboldt), 223–226 (Olber's paper "On the transparency of space").

31. Harrison, *Darkness at Night*, 110–112, 134, 141–145.

32. See Botting, *Humboldt*, 156, and Kellner, *Humboldt*, 82–83.

33. Kellner, *Humboldt*, 65.

Chapter 5 The Blue Flower

1. "Lines Composed a Few Miles above Tintern Abbey, on Revisiting the Banks of the Wye During a Tour. July 13, 1798," lines 95–102.

2. Cited in Richard Holmes, *Coleridge: Darker Reflections* (New York: Pantheon, 1998), 58. Consider also Johannes Brahms' *lied* "Feldeinsamkeit," Op. 86, No. 2: "I lie still in the high green grass / And gaze raptly upwards, / Crickets chirp around unceasingly, / The blue of the sky weaves wondrously around."

3. Mary W. Shelley, *Frankenstein: or, The New Prometheus* (Philadelphia: Running Press, 1987), 111.

4. "Berçant notre infini dans le fini des mers" ("Rocking our infinite in the finite of the seas"), from his poem "Le voyage," line 4; "Le ciel est triste et beau, comme un grand reposoir," from "Harmonies du soir," line 8, both in Baudelaire's *Fleurs du mal.*

5. Novalis, *Philosophical Writings,* tr. Margaret Mahony Stoljar (Albany: State University of New York Press, 1997); the quote is from Gage, *Color and Meaning,* 185. See the helpful essay by John Gage, "Mood Indigo—From the Blue Flower to the Blue Rider," in his *Color and Meaning,* 185–195, and also Wolfgang Müller-Funk, *Die Farbe Blau: Untersuchungen zur Epistemologie des Romantischen* (Vienna: Verlag Turia + Kant, 2000), especially 16–25. For the larger context of romantic views of the sky, see W. J. Lillyman, "The Blue Sky: A Recurrent Symbol," *Comparative Literature* 21, 116–124 (1969).

6. See Michael Heidelberger, "*Naturphilosophie*" in *The Routledge Encyclopedia of Philosophy,* ed. Edward Craig (London: Routledge, 1998), 6:737–743, and the helpful survey in Kenneth L. Caneva, "Physics and *Naturphilosophie*: A Reconnaissance," *History of Science* 35, 35–107 (1997).

7. Goethe, *Maxims and Reflections,* 115: "Die Natur verstummt auf der Folter: ihre treue Antwort auf redliche Frage ist: Ja! ja! Nein! nein! Alles übrige ist vom übel." Johann Wolfgang von Goethe, *Scientific Studies,* ed. tr. Douglas Miller (New York: Suhrkamp Publishers, 1988), 307. See also Erich Heller, "Goethe and Scientific Truth," in his book *The Disinherited Mind* (New York: Harcourt Brace Jovanovich, 1975), 3–34, especially 22–34 on the *Theory of Colors,* which gives the quote about the "torture chamber" at 22; and also my general discussion of the "torture of nature" in *Labyrinth,* 21–28.

8. "Freunde, flieht die dunkle Kammer," written in Weimar, 1823–1828, in *Goethes Gedichte in Zeitlicher Folge* (Frankfurt am Main: Insel, 1990), 1132 (my translation).

9. See G. W. Muncke, "Ueber subjective Farben und gefärbte Schatten," *Journal für Chemie und Physik* 30, 74–88 (1820) and Hoeppe, *Blau,* 91–92. On Goethe's color theory, see Dennis L. Sepper, *Goethe Contra Newton: Polemics and the Project for a New Science of Color* (Cambridge: Cambridge University Press, 1988). For a survey of the merits and faults of Goethe's work, see Hermann von Helmholtz, "Goethe's Scientific Researches" in his *Popular Scientific Lectures* (New York: Dover, 1962), 1–21, and Rudolf Magnus, *Goethe as a Scientist* (New York: Collier, 1961), 100–150, especially 111 (optical illusion is optical truth), 131 (on Goethe's explanation of the blue sky), 136 (Goethe's poem against Newton's optics). Other quotes come from 65 (smoke), 64 (mountains and also nearby objects shadowed blue by vapors). See also Brusatin, *History of Colors,* 102–114.

10. Johann Wolfgang von Goethe, *Theory of Colours,* tr. Charles Lock Eastlake (Cambridge, Mass.: MIT Press, 1970), 33–34. See also Michael Churma, "Blue Shadows: Physical, Physiological, and Psychological Causes," *Applied Optics* 33, 4719–4722 (1994).

11. Goethe, *Theory of Colours,* 68–71.

12. Johann Wolfgang von Goethe, *Scientific Studies,* ed. tr. Douglas Miller (New York: Suhrkamp, 1988), 151. See also Goethe, *Theory of Colours,* 60–68, and Sepper, *Goethe Contra Newton,* 149–150.

13. See his "Metamorphosis of Plants" in his *Scientific Studies,* 76–97.

14. Goethe, *Scientific Studies,* 151; Goethe, *Theory of Colours,* 64.

15. Goethe, *Scientific Studies,* 307.

16. Sepper, *Goethe Contra Newton,* 88.

17. Gage, *Color and Culture,* 191–212.

18. Goethe, *Theory of Colours,* 276.

19. Ibid., 310–311.

20. For Ritter and ultraviolet light, see Ronchi, *Nature of Light*, 275. For Ørsted, see Bern Dibner, *Oersted and the Discovery of Electromagnetism* (New York: Blaisdell, 1962).

21. See the excerpts in Magie, *Source Book in Physics*, 308–315, and Ronchi, *Nature of Light*, 237–241. For Young's demonstration of interference, see his "Experiments and Calculations relative to physical Optics," *Philosophical Transactions of the Royal Society* 1–16 (1804).

22. Thomas Young, "On the Theory of Light and Colors," in *Sources of Color Science*, ed. David L. MacAdam (Cambridge, Mass.: MIT Press, 1970), 51.

23. Christaan Huygens, *Treatise on Light*, tr. Silvanus P. Thompson (New York: Dover, 1962), 18–20.

24. William Shakespeare, Sonnet 60, lines 1–4.

25. See Henry John Steffens, *The Development of Newtonian Optics in England* (New York: Science History Publications, 1977), 107–136 (Young).

26. For Fresnel, Arago, Poisson, and Foucault, see Ronchi, *Nature of Light*, 241–259, and Jed Z. Buchwald, "Fresnel and Diffraction Theory," *Archives of International History of Science* 33, 36–111 (1983) and *The Rise of the Wave Theory of Light: Optical Theory and Experiment in the Early Nineteenth Century* (Chicago: University of Chicago Press, 1989).

27. Magie, *Source Book*, 280–283. For a helpful survey, see William A. Shurcliff and Stanley S. Ballard, *Polarized Light* (Princeton: Van Nostrand, 1964); for a more technical account, see William A. Shurcliff, *Polarized Light* (Cambridge, Mass.: Harvard University Press, 1966). See also G. P. Können, *Polarized Light in Nature*, tr. G. A. Beerling (Cambridge: Cambridge University Press, 1985), 11–28, 29–45 (blue sky and the clouds), 92–99 (natural scenery). For home experiments, see Bohren, *Clouds in a Glass of Beer*, 144–154, and *What Light through Yonder Window Breaks?*, 25–48.

28. Huygens, *Treatise on Light*, 52–105.

29. Ronchi, *Nature of Light*, 151, 255.

30. Magie, *Source Book*, 315–318.

31. See Hey, "From Leonardo to the Graser," I, 16. Let light come from a medium with index of refraction μ and reflect with an angle θ to the normal of the plane surface of a material with index of refraction μ', then at "Brewster's angle θ_B," defined by tan $\theta_B = \mu'/\mu$, the reflected light is completely polarized. For Brewster as the "last champion" of Newtonian optics, see Steffens, *Newtonian Optics in England*, 137–149.

32. Magie, *Source Book*, 324–334. For Humboldt's survey of Arago's work, see his preface to François Arago, *Meterological Essays* (London: Longman, Brown, 1855).

33. See Shurcliff and Ballard, *Polarized Light*, 95–97, and Können, *Polarized Light in Nature*, 11–12; for discussion of the physiology of this puzzling effect, see R. A. Bone and J. T. Landrum, "Dichroism of Lutein: A Possible Basis for Haidinger's Brushes," *Applied Optics* 22, 775 (1983).

34. *Oeuvres completes des François Arago* (Paris, 1858), 7:437–446, 10:277–281.

35. Sir John Leslie, *Treatises on Various Subjects of Natural and Chemical Philosophy* (Edinburgh: Adam and Charles Black, 1838), 496–497; Leslie often mentions Saussure and treats the cyanometer on 494–499. This work was published after Leslie's death in 1832; his original treatment can be found in his *Description of Instruments Designed for Extending and Improving Meteorological Observations* (Edinburgh: Abernethy and Walker, 1820). Middleton, "Random Reflections," 97, notes that Leslie believed that "the blue of the sky was due to the clear air itself," but Leslie's description of the sky as "coloured" does not indicate that he understood this as a result of scattering, as Middleton implies.

36. J. D. Forbes, *Transactions of the Royal Society of Edinburgh* 14, 371–374, 375–391 (1840), Middleton, "Random Reflections," 97, and Pedro Lilienfeld, "A Blue Sky History," *Optics & Photonic News* 32–39 (June 2004).

37. R. Clausius, "I. Ueber die Natur derjenigen Bestandtheile der Erdatmosphäre, durch welche die Lichtreflexion in derselben bewirkt wird," "II. Ueber die Blaue Farbe des Himmels und die Morgen- und Abenröte," *Annalen der Physik und Chemie* 76, 161–188, 188–195 (1849), discussed

by Elizabeth Garber, "Rudolf Clausius' Work in Meteorological Optics," *Rete* 2, 323–337 (1975), which gives a complete list of Clausius' publications on this question and a very helpful overview. For a treatment of his pioneering use of probability techniques in these studies, see Ivo Schneider, "Clausius' erste Anwendung der Wahrscheinlichkeitsrechnung im Rahmen der atmosphärische Lichtstreuung," *Archive for the History of Exact Science* 14, 143–158 (1974). Clausius' work became available in English in John Tyndall's translations, "On the Nature of those Constituents of the Atmosphere by which the Reflexion of the Light within it is effected" and "On the Blue Colour of the Sky and the Morning and Evening Red," published in *Scientific Memoirs Selected from the Transactions of Foreign Academies of Science*, ed. John Tyndall and William Francis (London: Taylor and Francis, 1853; republished, New York: Johnson Reprint, 1966), 303–325, 326–331.

38. E. Brücke, "Ueber die Farben, welche trübe Medien im auffallenden und durchfallenden Lichte zeigen," *Annalen der Physik und Chemie* 58, 363–385 (1853).

39. For the history of spectroscopy, see W. McGucken, *Nineteenth Century Spectroscopy* (Baltimore: Johns Hopkins University Press, 1969), and J. B. Hearnshaw, *The Analysis of Starlight: One Hundred and Fifty Years of Astronomical Spectroscopy* (Cambridge: Cambridge University Press, 1986).

40. G. Govi, "De la polarization de la lumière par diffusion," *Comptes rendues hebdomadaires des séances de l'Académie des sciences* 51, 360–361, 669–670 (1860).

41. Ibid., 361: though he does not mention Bouguer by name, Govi stresses that "light polarized by diffusion does not appear to have come from simple reflection from gas molecules."

Chapter 6 True Blue

I would like to thank Sarah Dry, Frank A. J. L. James, Francis O'Gorman, Jonathan Smith, and Maria Yamalidou for their helpful advice, criticisms, and suggestions concerning the material in this chapter regarding Tyndall and Ruskin.

1. John Tyndall includes these excerpts from Herschel's letter in his seminal paper, "On the Blue Colour of the Sky," 388–389. Herschel's other remarks about the sky's light come from John F. W. Herschel, *Meteorology* (Edinburgh: Adam and Charles Black, 1862), 230–231. Tyndall gives a nice overview of polarization phenomena in his *Familiar Lectures on Scientific Subjects* (London: Alexander Strahan, 1867), 342–399, and his *Notes of a Series of Nine Lectures on Light Delivered at the Royal Institution of Great Britain, April 8–June 3, 1869* (London: Longmans, Green, 1870), 57–71; I thank Stewart Greenfield for generously giving me a copy of this rare volume.

2. Tyndall, "On the Blue Colour of the Sky," 385. For a good overview of his work, see Hey, "From Leonardo to the Graser," I, 21–27. For biographical information, see A. S. Eve and C. H. Creasey, *Life and Work of John Tyndall* (London: Macmillan, 1945).

3. J. F. W. Herschel, "On a Case of Superficial Colour Presented by a Homogeneous Liquid Initially Colourless," *Philosophical Transactions of the Royal Society of London* Part I, 143–145 (1845); see also Hey, "From Leonardo to the Graser," I, 17–18.

4. Henry E. Roscoe, "On the Opalescence of the Atmosphere," *Proceedings of the Royal Institation*, 4, 651–659 (1866), at 658. See also Stokes' comment in appendix A, p. 191.

5. Tyndall, "On the Blue Colour of the Sky," 393. Helmholtz summarizes his work in "The Recent Progress of The Theory of Vision" in his *Popular Scientific Lectures*, 93–185; see also *Sources of Color Science*, 84–100.

6. George Gabriel Stokes: "On the Change of Refrangibility of Light," (1852), *Mathematical and Physical Papers by Sir George Gabriel Stokes* (Cambridge University Press, 1922), 3:259–413. Reviewing this argument, however, Lord Kelvin remarked that in conversation Stokes "did not tell me (though I have no doubt he knew it himself) why the light of the cloudless sky seen in any direction is blue, or I should certainly have remembered"; see Lord Kelvin, *Baltimore Lectures on Molecular Dynamics and the Wave Theory of Light* (Cambridge: Cambridge University Press, 1904), 302. Though he is not explicit, Stokes seems to have

reasoned that fluorescent light is *unpolarized* and due to molecular vibrations, hence *polarized* sky light cannot be due to molecules, but to larger particles. To unravel this problem required the advent of the quantum theory, which could address the inner workings of molecular vibrations. See Hey, "From Leonardo to the Graser," I, 16–21.

7. The quotes from Tyndall's notebooks are given by Andrew T. Young, "Rayleigh Scattering," *Physics Today* 35(1), 42–48 (1982), at 53.

8. See Tyndall, "On the Blue Colour of the Sky," and his fuller and more popular presentations of his experiments on sky blue (based on his Royal Institution lecture of January 15, 1869, which Ruskin read), "On Chemical Rays, and the Light of the Sky," *Fortnightly Review* 11, 226–248 (1869), and "On Chemical Rays and the Structure and Light of the Sky," in John Tyndall, *Fragments of Science for Unscientific People* (New York: Appleton, 1871), 235–273. Note that the photochemical "white clouds" Tyndall saw were probably composed of particles of order 100 nm in size, one hundred times smaller than the droplets (or ice particles) in atmospheric clouds (1000 nm). For a helpful guide to reproducing these phenomena, see E-Qing Zhu and Se-yuen Mak, "Demonstrating Colors of Sky and Sunset," *Physics Teacher* 32, 420–421 (1994), which also discusses Tyndall scattering in opal stones.

9. John Tyndall, "On a New Series of Chemical Reactions Produced by Light," *Proceedings of the Royal Society of London* 17, 92–102 (1868). See also John Tyndall, *Six Lectures on Light* (London: Longmans, Green, 1873), 152–161, at 157.

10. Tyndall, *Six Lectures on Light,* 156.

11. Tyndall, "On the Blue Colour of the Sky," 387.

12. John Ruskin, *Modern Painters: Of General Principles and of Truth,* vol. 1, in *Complete Works of John Ruskin,* 1:343–357, at 343, 346. See also Paul Fussell, *The Great War and Modern Memory* (London: Oxford University Press, 1975), 52–54.

13. *The Queen of the Air* in *Complete Works of John Ruskin,* 19:292–293, at 292.

14. *The Storm-Cloud of the Nineteenth Century*, in *Complete Works of John Ruskin*, 34:21–23, at 51, 53, 58.

15. Ibid., 52.

16. *Modern Painters*, vol. 1, in *Complete Works of John Ruskin*, 1:346.

17. Ibid., 1:347–349.

18. *The Queen of the Air*, in *Complete Works of John Ruskin*, 19:292–293 at 292. Tyndall responds in a beautiful memoir about his Alpine observations, "Climbing in Search of the Sky," *Fortnightly Review* 37 (new series), 1–15 (1870), at 4.

19. See Paul L. Sawyer, "Ruskin and Tyndall: The Poetry of Matter and the Poetry of Spirit," in *Victorian Science and Victorian Values: Literary Perspectives*, ed. James Paradis and Thomas Postlewait (New Brunswick, N.J.: Rutgers University Press, 1985), 217–246, quoting Ruskin on Saussure (221). See also Edward Alexander, "Ruskin and Science," *Modern Language Review* 64, 508–521 (1969), who notes that "his intense awareness of the new claims of science is perhaps most apparent in passages where he professes indifference to them" (508). See also Robert Hewison, "'Paradise Lost': Ruskin and Science," in *Time and Tide: Ruskin and Science*, ed. Michael Wheeler (London: Pilkington Press, 1996), 29–44, and Francis O'Gorman, "Ruskin's Science of the 1870s: Science, Education, and the Nation," in *Ruskin and the Dawn of the Modern*, ed. Dinah Burch (Oxford: Oxford University Press, 1999), 35–55. For Ruskin's ambition to be the president of the Geological Society, see Francis O'Gorman, *John Ruskin* (Phoenix Mill, Gloucestershire: Sutton Publishing, 1999), 6. For Ruskin and Saussure, see also Hewison, *John Ruskin*, 19–22. David Robertson notes that Saussure's writings were the best-known sources on the Alps among English readers; see "Mid-Victorians amongst the Alps," in *Nature and the Victorian Imagination*, ed. U. C. Knoepflmacher and G. B. Tennyson (Berkeley: University of California Press, 1997), 113–136, at 116. For Ruskin and Humboldt, see Paul Wilson, "'Over Yonder Are the Andes': Reading Ruskin Reading Humboldt," in *Time and Tide*, 65–84. See also Frederick Kirchhoff, "A Science against Sciences: Ruskin's Floral Mythology," 246–258, and Sharon Aronofsky Weltman, "Myth and Gender in Ruskin's Science," in *Ruskin*

and the Dawn of the Modern, 153–173. For Ruskin and the cyanometer, see Fitch, *Poison Sky*, 397–398, and Timothy Hilton, *The Pre-Raphaelites* (New York: Oxford University Press, 1970), 13.

20. John Tyndall, *Essays on the Use and Limit of the Imagination in Science* (London: Longmans, 1870 [second edition]), 54. See also Tyndall's essays entitled "Scope and Limit of Scientific Materialism," 107–124, and "Scientific Use of the Imagination," 125–166, in his *Fragments of Science*. For an insightful presentation of Tyndall's ideas in their context, see Jonathan Smith, *Fact and Feeling: Baconian Science and the Nineteenth-Century Literary Imagination* (Madison: University of Wisconsin Press, 1994), 34–37. On the Ruskin/Tyndall controversy, see especially Francis O'Gorman, " 'The Eagle and the Whale?' John Ruskin's Argument with John Tyndall," in *Time and Tide*, 45–64; for Ruskin's marginalia, see "Some Ruskin Annotations of John Tyndall," *Notes and Queries* 44, 348–349 (1997).

21. *The Queen of the Air*, in *Complete Works of John Ruskin*, 19:292–293, at 293–294. For their controversy about glacial movement, see the excellent discussion by Sawyer, "Ruskin and Tyndall," and also the helpful treatment of the context of Ruskin's sensibility, in Smith, *Fact and Feeling*, 152–179, which discusses Ruskin's letter to Charles Eliot Norton about putting the sky back in the bottle (173), found in *The Correspondence of John Ruskin and Charles Eliot Norton*, ed. John Lewis Bradley and Ian Ousby (Cambridge: Cambridge University Press, 1987), 135. See also the thorough discussion in Fitch, *Poison Sky*, 532–575.

22. *Storm-Cloud of the Nineteenth Century* in *Complete Works of John Ruskin*, 34:32–33. See Fitch, *Poison Sky*, 2–46, and the essays in *Ruskin and Environment: The Storm-Cloud of the Nineteenth Century*, ed. Michael Wheeler (Manchester: Manchester University Press, 1995), especially David Carroll, "Pollution, Defilement and the Art of Decomposition," 58–75. See also Martin A. Danahay, "Matter Out of Place: The Politics of Pollution in Ruskin and Turner," *Clio* 21, 61–77 (1991). For the influence of Krakatau (Krakatoa), see also Thomas A. Zaniello, "The Spectacular English Sunsets of the 1880s," in *Victorian Science and Victorian Values*, 247–267.

23. For Lodge's presentation of his version of the dust theory, see Oliver J. Lodge, "Dust," *Nature* 31, 265–269 (1884–85). His correspondence with Ruskin is found in *Complete Works of John Ruskin*, 37:514–527, at 520–521, 527. I am indebted to Smith, *Fact and Feeling*, 174–176, for an insightful discussion that drew this correspondence to my attention.

24. See Sawyer, "Ruskin and Tyndall," 241. Tyndall's Belfast address (1874) included atomism as part of a larger defense of science against theism; see Alan Willard Brown, *The Metaphysical Society: Victorian Minds in Crisis, 1869–1880* (New York: Columbia University Press, 1947), 231–238, and Ruth Barton, "John Tyndall, Pantheist. A Rereading of the Belfast Address," *Osiris* 3, 111–134 (1987). Regarding Tyndall's relation to atomic theory, see Sharon Mayer Libera, "John Tyndall and Tennyson's 'Lucretius,'" *Victorian Newsletter* 45, 19–22 (1974); Shigeo Sugiyama, "The Significance of the Particulate Conception of Matter in John Tyndall's Physical Researches," *Historia Scientiarum* 2, 119–138 (1992); and Maria Yamalidou, "John Tyndall, the Rhetorician of Molecularity. Part One. Crossing the Boundary towards the Invisible. Part Two. Questions Put to Nature," *Notes and Records of the Royal Society (London)* 53, 231–242, 319–331 (1999).

25. See Tyndall's 1876 address on "Fermentation, and its Bearings on Surgery and Medicine" and his 1878 essay on "Spontaneous Generation," *Fragments of Science*, pp. 251–289, 290–334, as well as "On Dust and Disease" in his *Essays on the Floating-Matter of the Air in relation to Putrefaction and Infection* (New York: Appleton, 1882; reprint, New York: Johnson Reprint, 1966), 1–43. For a fine discussion of the larger context, see James Edgar Strick, *Sparks of Life: Darwinism and the Victorian Debates over Spontaneous Generation* (Cambridge, Mass.: Harvard University Press, 2000), especially 35–61 (on the "molecular" theories of histology), 134–182 (Tyndall's role in the controversy).

26. Tyndall, *Fragments of Science*, 121 (sky matter), 268 (floating dust of the air). In 1870, the *Lancet* concluded that "his lecture was a very skilful attempt to familiarise the public mind with the existence of atmospheric particles, and to lead up to and encourage, without absolutely expressing the idea that germs are particles, and that particles may be germs"

causing the blue of the sky. Cited in Tyndall, *Essays on the Use and Limit of the Imagination in Science,* 8. Tyndall also connected the greenness of the sea with suspended particles; see his "From Portsmouth to Ovan to See the Eclipse," *Fortnightly Review* 15, 330–351 (1871), at 346–351.

27. See appendix B, 189–191 (Stokes on "molecular reflection"), 194 (Tyndall on size of emergent particles).

28. As stated in his 1880 essay on "Goethe's '*Farbenlehre,*'" in John Tyndall, *New Fragments* (New York: Appleton, 1897), 47–75, at 61.

29. See appendix B, 189–191 (Tyndall vs. Stokes on the possibility of molecular reflection).

30. Tyndall, "On the Blue Colour of the Sky," 388, also notes that the problem of scattering "*in* air *upon* air" had "led many of our eminent men, Brewster himself amongst their number, to entertain the idea of *molecular reflection,*" showing the persistence of Bougner's idea in the 1830s. I have not found any such reference in Brewster's works. Tyndall's context suggests that he is here thinking of a molecular origin for Brewster's law of polarization, which he goes on to discuss. For a treatment of the experimental verification of these phenomena, see Wallace A. Hilton, "An Experiment on Sky Polarization and Brightness," *Physics Teacher* 16, 294–296 (1978).

31. For a discussion that clarifies the fundamental identity between these seemingly different phenomena, see William T. Doyle, "Scattering Approach to Fresnel's Equations and Brewster's Law," *American Journal of Physics* 53, 463–468 (1985).

32. Tyndall, *Six Lectures on Light,* 153–154. He amplifies his treatment of clouds in his "Note on the Formation and Phenomena of Clouds," *Philosophical Magazine* 38:156–158 (1869).

33. In a letter dated January 27, [1869?], M. F. Egerton asked Tyndall: "since the colour seems to depend on the attenuation of the vapour fitting the more rapid vibrations, why I wonder do we never get a violet, even lavender sky in dry climates? Do you get it in the tube?" (Royal Institution of Great Britain JT/1/E/13).

34. John Tyndall, *Faraday as a Discoverer* (New York: Thomas Y. Crowell, 1961).

35. Faraday's sole statement on the question of the blue sky is found in a letter to C. R. Leslie of May 25, 1854, in *The Selected Correspondence of Michael Faraday*, ed. L. Pearce Williams (Cambridge: Cambridge University Press, 1971), 2:735–736, responding to Leslie's question about how to explain the sky's blue in his forthcoming *A Hand-Book for Young Painters*. There is no evidence Faraday was at all aware of Brücke's 1853 paper, to which I have found no reference in Faraday's writings. Faraday's claim that he had confirmed Saussure's direct observation of "cloud-vesicles" is recorded by Tyndall "as I learnt from himself" in "The Sky," *Fragments of Science*, 1:131–141, at 136. I thank Frank A. J. L. James for helpful information about Faraday's involvement in this question.

36. For Maxwell's color top, see C. W. F. Everitt, *James Clerk Maxwell: Physicist and Natural Philosopher* (New York: Scribner's Sons, 1975), 63–67; for his work on color theory, see his essay "On Color Vision," in *Sources of Color Science*, 75–83, and P. M. Harman, *The Natural Philosophy of James Clerk Maxwell* (Cambridge: Cambridge University Press, 1998), 37–48. For Rayleigh's reaction, see Robert John Strutt, Fourth Baron Rayleigh, *Life of John William Strutt, Third Baron Rayleigh* (Madison: University of Wisconsin Press, 1968), 51–54.

37. Rayleigh's first arguments are found in his 1871 paper "On the Light from the Sky, Its Polarization and Colour," in *Scientific Papers by Lord Rayleigh* (New York: Dover, 1964), 1:87–103. An excerpt from this paper is also available in a helpful anthology that has an introduction and commentary: Robert Bruce Lindsay, *Lord Rayleigh—The Man and His Work* (Oxford: Pergamon Press, 1970), 85–92. In this paper, Rayleigh assumes that the incoming wave is sinusoidal and calculates its acceleration. Assuming the mass density of the particle, he then can write the force on that particle in terms of that acceleration and its density, using Newton's third law. Note that in 1871 he elaborates this mathematical treatment in another paper, "On the Scattering of Light by Small Particles," *Scientific Papers by Lord Rayleigh* 1:104–110. See also the excel-

lent review article by Young, "Rayleigh Scattering," which is very helpful in clarifying the confusing variety of uses of the phrase "Rayleigh scattering" in the physics literature; Hey, "From Leonardo to the Graser," II; the mathematical tour de force of S. Chandrasekhar and Donna D. Elbert, "The Illumination and Polarization of the Sunlit Sky: On Rayleigh Scattering," *Transactions of the American Philosophical Society* 44, 643–654 (1954) (included in Bohren, *Scattering,* 261–272); and G. V. Rozenberg, "Light Scattering in the Earth's Atmosphere," *Soviet Physics Uspekhi* 3, 346–371 (1960) (also in Bohren, *Scattering,* 88–113).

38. Maxwell introduced dimensional analysis in his paper "On the Elementary Relations of Electrical Quantities" (with Fleeming Jenkin), *Report of the British Association for the Advancement of Science,* 1st. ser., 32, 130–163 (1863). For a nice set of applications, see Craig F. Bohren, "Dimensional Analysis, Falling Bodies, and the Fine Art of *Not* Solving Differential Equations," *American Journal of Physics* 72, 534–537 (2004).

39. At the beginning of "On the Light from the Sky" (*Scientific Papers by Lord Rayleigh* 1:90), Rayleigh notes that "the principal result may be anticipated from a consideration of the *dimensions* of the quantities concerned." He begins by noting that the brightness of the scattered light is proportional to the brightness of the incident light; we are seeking their ratio, which must therefore be dimensionless, whatever the dimensions of brightness. (Brightness—the proper technical term is irradiance—is defined as the energy received per m^2.) We assume that the scattering particles are so small compared to the incoming wavelength that we can treat each particle as reacting instantaneously as a whole to the incident wave. Thus, the ratio will depend on the volume of each particle (call it V), rather than on its size or cross-sectional area. The other relevant factors that might enter are the wavelength of the light (λ) and the distance from the observer to the particle (r). The speed of light (c) or the mass of the particle (m) cannot enter in because they would introduce dimensions of mass or time that cannot appear in the dimensionless ratio. We assume that the ratio is inversely proportional to r^2, as is commonly true sufficiently far from most light sources. Now the wavelength and volume of the particle must multiply or divide $1/r^2$

to give a dimensionless quantity. V has the dimensions of length cubed; r and λ have dimensions of length. Therefore the quantity $V^2/r^2\lambda^4$ has dimensions $(\text{length})^6/(\text{length})^2(\text{length})^4$, which means it is dimensionless. Since this is the only way this can be accomplished, the ratio must be proportional to $V^2/r^2\lambda^4$, inversely proportional to the fourth power of the wavelength, as Rayleigh claimed.

This can be further justified by noting that in general the brightness of a scattered wave (and hence also our ratio) is proportional to the square of its amplitude; in turn, that amplitude is proportional to the volume of the scatterer. Therefore the ratio is proportional to V^2; we already know the ratio is proportional to r^{-2} and so the only way that the wavelength λ can enter to make this dimensionless is as λ^{-4}.

40. *Scientific Papers by Lord Rayleigh* 1:95. In this case, the ratio depends on only two quantities, the wavelength λ and δ, the thickness of the film. Since for purposes of approximation we treat the film as very large, we do not expect dependence on r since the film looks the same from any distance. Then the only way to form a dimensionless amplitude is δ/λ (since the reflected amplitude varies as δ, from Newton's treatment) and hence the ratio is proportional to δ^2/λ^2, inversely as the square of the wavelength.

41. "On the Light from the Sky," *Scientific Papers by Lord Rayleigh* 1:93–94, 102–103. I thank John Howard, curator of the Rayleigh Archive at the Air Force Research Laboratory (Bascom Air Force Base, Bedford, Mass.) as well as Jennifer Sullivan and Spencer Weart at the Niels Bohr Library, American Institute of Physics, for their kind help in making available Rayleigh's notebooks and papers for my study. For a useful overview, see John N. Howard, "The Rayleigh Notebooks," *Applied Optics* 3, 1129–1133 (1964). Rayleigh's observations covered only a very restricted part of the spectrum, about 486–656 nm, notably excluding the violet range. His original notes can be found in his experimental notebooks for 1870 on November 1–13 (box II, reel 3). See also Kelvin's discussion in his *Baltimore Lectures*, 319–320. I will return to modern measurements of the skylight spectrum in chapter 10.

42. "On the Light from the Sky," *Scientific Papers by Lord Rayleigh* 1:102.

43. For further discussion of these physical arguments, see Peter Pesic, "The Sky Is Falling: Newton's Droplets, Clausius's Bubbles, and Tyndall's 'Sky Matter,'" *European Journal of Physics* 26, 189–193 (2005).

Chapter 7 Blue Laws

1. See *The Scientific Letters and Papers of James Clerk Maxwell*, ed. P. M. Harman (Cambridge: Cambridge University Press, 1995), 2:617–619, at 619. For Tyndall's letter to Rayleigh, see Robert Strutt, *Life of John William Strutt*, 54.

2. *Scientific Letters and Papers of Maxwell*, 2:919–920. For reactions to Rayleigh's 1871 work, see Robert Strutt, *Life of John William Strutt*, 53–54. He had written another, shorter paper in 1871, "On the Scattering of Light by Small Particles," *Scientific Papers by Lord Rayleigh* 1:104–110, redoing his treatment within the framework of analytical mechanics and beginning to consider the possible nonspherical shape of the particles. See also the commentary by Victor F. Weisskopf, "Search for Simplicity: Maxwell, Rayleigh, and Mt. Everest," *American Journal of Physics* 54, 13 (1986), which errs in stating that Maxwell wrote Rayleigh from Darjeeling, which he never visited; his 1873 letter is dated from his Scottish estate.

3. For further details, see Pesic, "The Sky Is Falling."

4. Tyndall, "Scientific Use of the Imagination," *Fragments of Science*, 2:108. For the larger background, see *The Atomic Debates*, ed. W. H. Brock (Leicester: Leicester University Press, 1967), 1–11 (on Davy and John Herschel), and Mary Jo Nye, *Molecular Reality* (New York: Elsevier, 1972). For a valuable collection of original papers, see *The Question of the Atom*, ed. Mary Jo Nye (Los Angeles: Tomash, 1984).

5. See James Clerk Maxwell, *Theory of Heat*, ed. Peter Pesic (Mineola, N.Y.: Dover, 2001), xv–xix, 301–333, and the excellent collection *Maxwell on Molecules*, ed. Elizabeth Garber, Stephen G. Brush, and C. W. F. Everitt (Cambridge, Mass.: MIT Press, 1986).

6. *Scientific Letters and Papers of Maxwell*, 2:919–920.

7. The 1881 paper, "On the Electromagnetic Theory of Light," *Scientific Papers by Lord Rayleigh* 1:518–536, uses Maxwell's equations and reaffirms the basic conclusion of the 1871 argument (526). These same results are confirmed in his 1899 paper, "On the Transmission of Light through an Atmosphere Containing Small Particles in Suspension, and on the Origin of the Blue of the Sky," *Scientific Papers by Lord Rayleigh* 4:398–405 (in Bohren, *Scattering,* 26–34). Rayleigh returned to these issues again in a different context in his 1910 paper entitled "Colours of Sea and Sky," *Scientific Papers by Lord Rayleigh* 5:540–546. He also gave a wonderful survey in his entry "Sky" for the *Encyclopaedia Brittanica,* 11th ed. (New York: Encyclopaedia Brittanica, 1911), 26:202–205. For the relation of Rayleigh's work to the modern optical theorem, see Roger G. Newton, "Optical Theorem and Beyond," *American Journal of Physics* 44, 639–642 (1976).

8. Lorenz's work remained little-known even after it was republished in French as "Sur la lumière réfléchie et réfractée par une sphère transparente" in *Œuvres Scientifiques de L. Lorenz,* ed. H. Valentiner (Copenhagen: Librarie Lehmann & Stage, 1898; reprint, New York: Johnson Reprint, 1985), 1:405–502. See Helge Kragh, "Ludvig Lorenz and Nineteenth Century Optical Theory: The Work of a Great Danish Scientist," *Applied Optics* 30, 4688–4695 (1991), and Lilienfeld, "A Blue Sky History." Ludvig Lorenz should not be confused with the Dutch scientist H. A. Lorentz, who also did important work relating the refractive index to the dielectric constant of a medium; both are remembered in the Lorentz–Lorenz equation. See H. A. Lorentz, *Problems of Modern Physics,* ed. H. Bateman (Boston: Ginn and Company, 1927), 52–60.

9. See J. Gay-Lussac, "Memoir on the Combination of Gaseous Substances with each other," *Mémoires de la Société d'Arcueil* 2, 207–234 (1809), and A. Avogadro, "Essay on a Manner of Determining the Relative masses of the Elementary Molecules of Bodies, and the Proportions in which they enter into their Compounds," *Journal de Physique* 78, 58–76 (1811), as translated in *Foundations of the Molecular Theory* (Edinburgh: Alembic Club, 1961).

10. See Alfred Bader and Leonard Parker, "Joseph Loschmidt, Physicist and Chemist," *Physics Today* 54(3), 45–50 (2001). For Thomson's 1870

paper entitled "Voltaic Potential Differences and Atomic Sizes," see Sir William Thomson, Baron Kelvin, *Mathematical and Physical* Papers (Cambridge: Cambridge University Press, 1911), 5:284–296. For H. A. Lorentz's discussion of this issue, see his *Problems of Modern Physics*, 60.

11. See Maxwell's 1873 paper, "On Loschmidt's Experiments in Diffusion in relation to the Kinetic Theory of Gases," in *The Scientific Papers of James Clerk Maxwell*, ed. W. D. Niven (New York: Dover, 1965), 2:342–350.

12. Kelvin's detailed discussion and calculations of Avogadro's number are given in his *Baltimore Lectures*, 279–323. He had begun to work on atomic sizes long before; in a letter to Stokes of January 7, 1883, he asked whether Tyndall's particles are "demonstrably a small fraction of λ, in their diameters?" See *The Correspondence between Sir George Gabriel Stokes and Sir William Thomson, Baron Kelvin of Largs*, ed. David B. Wilson (Cambridge: Cambridge University Press, 1990), 2:536–537, and also the excellent treatment in David B. Wilson, *Kelvin and Stokes: A Comparative Study in Victorian Physics* (Bristol: A. Hilger, 1987).

13. See Bouguer's *Optical Treatise*, 235–239, and W. E. Knowles Middleton, "Bouguer, Lambert, and the Theory of Horizontal Visibility," *Isis* 51, 145–148 (1960), and his *Vision through the Atmosphere* (Toronto: University of Toronto Press, 1952). See also Bohren, *Clouds in a Glass of Beer*, 120–127.

14. *Scientific Papers by Lord Rayleigh* 4:403. Kelvin records his observations in his 1899 letter on the "Blue Ray of Sunrise over Mont Blanc," *Mathematical and Physical Papers* 5:231, noting that Jules Verne's celebrated "green flash" (*rayon vent*) "is the corresponding phenomenon at sunset; which I first saw about six years ago."

15. For more complete details, see Peter Pesic, "Estimating Avogadro's Number from Skylight and Airlight," *European Journal of Physics* 26, 183–187 (2005). Light traveling through the atmosphere is extinguished by a factor of $1/e$ after traveling a distance $1/\beta$ (where e is the base of natural logarithms, $e = 2.718\ldots$). Rayleigh's formula relates the extinction coefficient β, the number of scattering molecules per cc n, the wavelength λ, and the index of refraction of air $\mu = 1.0003$ so that $n = (32\pi^3(\mu$

$- 1)^2)/(3\beta\lambda^4)$. Using a standard criterion for the visual range R_v, it turns out that $R_v \approx 3.9/\beta$. Then in the visible range $\lambda = 400$–$700\,\text{nm}$, $n = (3.2 \times 10^{20})/\beta$ (if β is in m^{-1}). Since one mole at standard temperature and pressure (0°C and one atmosphere) occupies 22.4 liters $= 2.24 \times 10^4\,\text{cc}$, then $N = (2.24 \times 10^4)n$ and Rayleigh's formula leads to $N = (7.3 \times 10^{18})/\beta = (1.8 \times 10^{18})R_v$, where distances are measured in m. Thus, my value for the visual range in Sante Fe, 160 km, leads to a value of roughly 3×10^{23} for Avogadro's number.

16. For fuller details, see Pesic, "Estimating Avogadro's Number." Using a cardboard mailing tube, you can compare the luminance of airlight in front of a distant dark object to that of skylight, dimmed and reflected in a black piece of glass. I used a nearby hill (about $d = 500\,\text{m}$ away) whose luminance seemed to match the reflected sky luminance. It turns out that a fraction $r = 0.05$ of the light incident on black glass will reflect back. My paper shows that Avogadro's number is $N_A = 2rRT/dmg = (5.3 \times 10^{26})/d$, where m is the average molecular weight of air (28.94), T is room temperature (300 K), R the gas constant (8.3 J/mole–K), and g the acceleration of gravity ($9.8\,\text{m/sec}^2$). Then if $d = 500\,\text{m}$, this gives a value for Avogadro's number $N_A = 11 \times 10^{23}$.

17. See his *Life of John William Strutt*, 299–301. There is a valuable account in Hey, "From Leonardo to the Graser," part IV. For modern treatments of Rayleigh scattering and the color of the sky, see H. C. van de Hulst, "Scattering in the Atmosphere of the Earth and the Planets," and his classic book *Light Scattering by Small Particles* (New York: Dover, 1981), 63–84, 414–439; Milton Kerker, *The Scattering of Light and Other Electromagnetic Radiation* (New York; Academic, 1969), 1–3, 27–39, 574–583; Craig F. Bohren and Donald R. Huffman, *Absorption and Scattering of Light by Small Particles* (New York: Wiley, 1983); and Walter T. Grandy, Jr., *Scattering of Waves from Large Spheres* (Cambridge: Cambridge University Press, 2000), 1–29, 107–115.

18. Jean Cabannes, *La Diffusion Moléculaire de la Lumière* (Paris: Presses Universitaires de France, 1929), 9.

Chapter 8 Blue Riders

1. See Abraham Pais, *"Subtle is the Lord . . .": The Science and the Life of Albert Einstein* (New York: Oxford University Press, 1982), 79–107.

2. For a treatment of Einstein in his practical milieu, see Peter Galison, *Einstein's Clocks, Poincaré's Maps: Empires of Time* (London: Sceptre, 2003).

3. See Pais, *"Subtle is the Lord . . ."*, 93–100, and Nye, *Molecular Reality*, which also discusses the history of Jean Perrin's contribution. Einstein's original papers on Brownian motion are helpfully collected in Albert Einstein, *Investigations on the Theory of the Brownian Movement*, ed. R. Fürth (New York: Dover, 1956), especially Einstein's less technical account, "The Elementary Theory of the Brownian Motion," 68–85. For modern treatments, see Robert M. Mazo, *Brownian Motion: Fluctuations, Dynamics, and Applications* (Oxford: Clarendon Press, 2002), 1–10 (historical survey), 46–61 (Einstein–Smoluchowski theory). Brown's initial observations were not of pollen grains (which are too large to show this motion) but clay particles contained in pollen, as emphasized by David M. Wilkinson, "Brown Knew Particles Were Smaller Than Pollen," *Nature 434*, 137 (2005).

4. Jean Perrin, *Atoms* (Woodbridge, Conn.: Ox Bow Press, 1990), 206–208. This classic work is of great value and is wonderfully readable. See also his earlier treatment, *Brownian Movement and Molecular Reality*, tr. F. Soddy (London: Taylor and Francis, 1910).

5. Perrin's first quote comes from his *Brownian Movement*, 91; his second from his book *Atoms*, 207–208.

6. See H. Eugene Stanley, *Introduction to Phase Transitions and Critical Phenomena* (New York: Oxford University Press, 1971), 3–21. For the early history, see Hey, "From Leonardo to the Graser," I, 21. See also the treatment of experimental and theoretical aspects of opalescence by Y. Rocard in Cabannes, *Diffusion Moléculaire de la Lumière*, 265–295. The best way to see this phenomenon yourself involves a special compound, 2,6-lutidine (also called 2,6-dimethylpyridine), mixed with water. This (extremely unpleasant smelling) compound is available from chemical

supply houses and has a convenient critical temperature, 33.4°C. Once you have obtained it, the procedure for observing critical opalescence is given in W. I. Goldburg, "Dynamic Light Scattering," *American Journal of Physics* 67, 1152–1160, at 1154–1155 (1999). See also A. C. Mowery and D. T. Jacobs, "Undergraduate Experiment in Critical Phenomena: Light Scattering in a Binary Fluid Mixture," *American Journal of Physics* 51, 542–545 (1983).

7. Marian Smoluchowski, "Beitrag zur Theorie der Opaleszenz von Gasen im kritischen Zustande," *Bulletin de l'Academie des Sciences de Cracovie*, 493–502 (1911). The description in terms of colloids comes from W. H. Martin, "The Scattering of Light in One-Phase Systems," in *Colloid Chemistry*, ed. Jerome Alexander (New York: Chemical Catalogue, 1926), 1:346. Much earlier, Smoluchowski had worked on the pressure distribution in the atmosphere of the planets; see his paper "Über die Atmosphäre der Erde und der Planeten," *Physikalishe Zeitschrift* 2, 307–313 (1900), in *Oeuvres de Marie Smoluchowski*, ed. Ladislas Natanson and Jean Stock (Cracow: Imprimerie de l'Université Jaguellonne, 1924), 1:263–278. For his work on Brownian movment, see Bronislaw Średniawa, "The Collaboration of Marian Smoluchowski and Theodor Svedberg on the Brownian Motion and Density Fluctuations," *Centaurus* 35, 325–355 (1993). For an overview of his works and biography, see Armin Teske, "Einstein und Smoluchowski: Zur Geschichte der Brownschen Bewegung und der Opaleszenz," *Sudhoffs Archiv* 53, 292–305 (1969), and *Marian Smoluchowski: Leben und Werk* (Warsaw: Wydawnictwo Polskiej Akademii Nauk, 1977). See also two very helpful collections of essays and translations: *Marian Smoluchowski: His Life and Scientific Work*, ed. S. Chandrasekhar, M. Kac, and R. Smoluchowski (Warsaw: Polish Scientific Publishers, 2000), especially Roman Smoluchowski's biographical memoir about his father (9–14), from which I have drawn material about Smoluchowski's mountaineering, the nice surveys by Kac (15–20) and Chandrasekhar (21–28), and the translations of important papers of 1906 and 1916 by Smoluchowski; and *Essays Devoted to Scientific and Didactic Work of Marian Smoluchowski (1872–1917)*, ed. Bronisław Średniawa (Cracow: Nakladem Uniwersytetu Jagiellonskiego, 1991), especially Bronisław Średniawa, "Marian Smoluchowski's Collaboration with Experimentalists in the Investigations of Brownian Motion and Density

Fluctuations" (9–46) and K. Rościszewski, "The Impact of Smoluchowski's Works on the Process of the Birth of the Modern Statistical Physics" (47–68).

8. Smoluchowski, "On Opalescence of Gases in the Critical State," *Philosophical Magazine* 23, 165–173 (1912). Einstein's obituary of him appears in *The Collected Papers of Albert Einstein*, ed. Martin J. Klein, A. J. Kox, Jürgen Rein, and Robert Schulmann (Princeton: Princeton University Press, 1993), 6:576–579.

9. See Lorentz, *Problems of Modern Physics*, 55–58, and Richard P. Feynman, Robert B. Leighton, and Matthew Sands, *The Feynman Lectures on Physics* (Reading, Mass.: Addison-Wesley, 1963), 1:32–35; for the connection with fluctuations, see Francis A. Jenkins and Harvey E. White, *Fundamentals of Optics* (New York: McGraw-Hill, 1957 [third edition]), 458, and André Guinier, *The Structure of Matter from Blue Sky to Liquid Crystals* (Baltimore: Edward Arnold, 1984), 52–55, which also emphasizes the connection between atomic theory and the blue sky. For the comparison with fluids, see B. H. Zimm, "Molecular Theory of the Scattering of Light in Fluids," *Journal of Chemical Physics*, 13, 141 (1945). For simple estimates of the intermolecular spacing (mean free paths) in gases, see F. Reif, *Fundamentals of Statistical and Thermal Physics* (New York: McGraw-Hill, 1965), 471. Since the mean free path in a gas is inversely proportional to its pressure, we can estimate that when we are above an altitude such that the pressure is $1/e$ that at sea level, roughly 8 km, then the mean free path is more than e times larger than at sea level (300 nm), hence more than 800 nm. Thus, air molecules in the upper atmosphere scatter incoherently because they are distant, on the scale of visible wavelengths.

10. A. Einstein, "The Theory of the Opalescence of Homogeneous Fluids and Liquid Mixtures near the Critical State," *Annalen der Physik* 25, 205–226 (1910), as translated in *Collected Papers of Albert Einstein*, 3:286–312 (translation: 3:231–249). This edition also includes a very helpful note, "Einstein on Critical Opalescence," 3:283–285.

11. Einstein, "The Theory of the Opalescence of Homogeneous Fluids," *Collected Papers of Albert Einstein*, 3:245, n. 7 (translation).

12. These quotes from the Einstein–Smoluchowski correspondence are found in *Collected Papers of Albert Einstein*, 3:283–285.

13. This has been emphasized by Craig F. Bohren and Alistair B. Fraser, "Colors of the Sky," *Physics Teacher* 23, 267–272 (1985).

14. For a general treatment, see Peter Pesic, *Seeing Double: Shared Identities in Physics, Philosophy, and Literature* (Cambridge, Mass.: MIT Press, 2001), 87–99.

15. See Peter Pesic, "Estimating *hc*/*k* from Skylight," *American Journal of Physics* 73, 457–458 (2005), which uses the effective temperature of the sun as a way to determine *hc*/*k*, where *h* is Planck's constant, *c* the speed of light, and *k* is Boltzmann's constant. If one takes *c* and *k* as known and uses the visible range of light, 400–700 nm, this gives an estimate for $h = 5.5–9.2 \times 10^{-34}$ J-sec, the same order of magnitude as the standard value, 6.67×10^{-34} J-sec.

16. Smoluchowski, "Experimentelle Bestätigung der Rayleigh'schen Theorie des Himmelsblaus," *Bulletin de l'Academie des Sciences de Cracovie*, 218–220 (1916); for Einstein's obituary, see p. 235, note 8.

17. For Goethe and Luke Howard, see the discussions in Kurt Badt, *Constable's Clouds* (London: Routledge and Kegan Paul, 1950), 9–34; John Thornes, "Constable's Clouds," *Burlington Magazine* 121, 697–704 (1979); Lee and Fraser, *Rainbow Bridge*, 83–87; and Peter Galison, *Image and Logic: A Material Culture of Microphysics* (Chicago: University of Chicago Press, 1997), 75–80. For the influence of Krakatau (Krakatoa), see Zaniello, "Spectacular English Sunsets."

18. On clouds in a bottle, see Bohren, *Clouds in a Glass of Beer*, 1–14. For cloud chambers, see Galison, *Image and Logic*, 65–105, especially 85 (Krakatoa), 90 (quote from Wilson), 81–84 (illustrations of Wilson's cloud photography).

19. For directions on building a cloud chamber at home, see C. L. Stong, *The Amateur Scientist* (New York: Simon and Schuster, 1960), 307–334.

20. See Gage, *Color and Meaning*, 192–195 (which includes the quotes from Kandinsky), and Wassily Kandinsky, *Concerning the Spiritual in Art*,

tr. M. T. H. Sadler (New York: Dover, 1977), 36–38. See also Brusatin, *History of Colors*, 136–153, and Christoph Schreier, "Von 'inneren Klang' einer Farbe: Wassily Kandinsky und der Blaue Reiter," in *Blau: Farbe der Ferne*, 179–187. See also William Gass, *On Being Blue* (Boston: David R. Godine, 1991), especially 6–7, 74–77 (on Kandinsky), 82–86. For Wilhelm Reich's theories about sky blue as a manifestation of his "orgone energy," see his *The Cancer Biopathy* (New York: Farrar, Straus, and Giroux, 1973), 99–108, 120–122 (Albert Einstein's involvement in some experiments suggested by Reich).

21. Leo Tolstoy, *War and Peace*, tr. Louise and Aylmer Maude (New York: Knopf, 1992), vol. 1, 357–358 (Andrei on the battlefield), 222 (no sky but atmosphere), 500–501 (Pierre points to the sky).

22. The examples from Baudelaire, Mallarmé, and Stevens (and others) are found in Lillyman, "The Blue Sky," 122–124. See also Lois Oppenheim, "'Le Ciel est morte': Mallarmé and a Metaphysics of (Im)possibility," *Analecta Husserliana* 23, 177–188 (1988).

23. See Fussell, *The Great War*, 51–63, especially 54 (Max Plowman quote).

24. See Ruskin, *Modern Painters*, in *Collected Works of John Ruskin*, 1:315.

Chapter 9 Midnight Blue

1. Greenler, *Rainbows*, 127–128, contains excellent illustrations. Some typical time-exposure values for a clear sky with a gibbous moon are ISO 100 and *f*5.6 for about 15 minutes, but some experimentation is necessary. Here, a film camera tends to produce better-looking results than a digital camera. If you do try this with a digital camera, its noise suppression must be turned on and you should consult the camera's manual for instructions about taking time exposures. Note than many published astronomical photographs have had their background color altered to make them *look* black, though the actual color would have included airglow, scattered moonlight, or other colored background light. I thank David Malin for his expert advice on these matters. Robert John Strutt (fourth baron Rayleigh) recorded his observations of the color of the

night sky in July 1920 (paragraphs 1217–1223 of his notebooks for that year) and in February–March 1923 (paragraphs 1531, 1539–1540).

2. Greenler, *Rainbows*, 137, notes that it is sometimes possible to see Venus in daylight and speculates this might be the origin of the legend of stars visible up chimneys. David W. Hughes, "On Seeing Stars (Especially Up Chimneys)," *Quarterly Journal of the Royal Astronomical Society* 24, 246–257 (1983), calculates that it is just possible that planets or zeroth magnitude objects could be seen in daylight, if one knew exactly where to look; however, his calculations show that no such objects could be seen up a chimney so that "we have to conclude the whole legend is untenable." For an example of the persistence of this belief, consider Ford Madox Ford's novel *A Man Could Stand Up—*, in which a character fighting in the Great War muses on the paradox that "the light seen from the trench seemed, if not brighter, then more definite. So, from the bottom of a pit-shaft in broad day you can see the stars." See Fussell, *The Great War*, 52.

3. The essential reference for the puzzle of night sky darkness is Harrison, *Darkness at Night*, 75–90 (Halley and Chéseaux), 155–165 (Kelvin), 146–154 (Poe), 127–136 (cosmic hierarchy), 176–194 (cosmic expansion and redshift), 195–204 (energy limitations), 117–123 (the great debate). Harrison also treats these issues in the larger context of cosmology in his excellent text *Cosmology: The Science of the Universe*, second edition (Cambridge: Cambridge University Press, 2000), 515–534.

4. The quotation from *Eureka* is taken from Harrison, *Darkness at Night*, 148.

5. See Roach and Gordon, *The Light of the Night Sky*, which includes a wealth of very helpfully presented data (mean surface brightness of the night sky, 22–29; galactic and cosmic light, 83–88, 107–117). For an even more recent and authoritative compilation, see C. Leinert, S. Bowyer, et al., "The 1997 Reference of Diffuse Night Sky Brightness," *Astronomy and Astrophysics Supplement Series* 127, 1–99 (1998).

6. See Peter Pesic, "Brightness at Night," *American Journal of Physics* 66, 1013–1015 (1998), which I will summarize here: Let us use the common astronomical unit of distance, the parsec, which is 3.3 light-years (that

is, 3×10^{13} kilometers or 2×10^{13} miles). By the "number density of stars," n, we mean the number of stars per cubic parsec. Then Kelvin's calculation can be summarized in this way: The product of the number density of stars and their average lifetime (t, in years) is about 60 years per cubic parsec ($nt \sim 60$). From this relation, Kelvin's value for $n = 0.2$ yields $t = 300$ years, far too small. If we take the Andromeda galaxy (M31) as typical both of number of stars (10^{11}) and intergalactic spacing (1 million parsecs from our galaxy), then one gets a value for n of about 10^{-8} stars per cubic parsec, from which Kelvin's relation then predicts t of about one billion years (10^9), in rough agreement with the present value of about 14 billion years.

7. See Joe D. Burchfield, *Lord Kelvin and the Age of the Earth* (Chicago: University of Chicago Press, 1990), especially 134–162. For a helpful survey of the cosmology of Kelvin's time and its connection with the geological age problem, see Harrison, *Cosmology*, 66–86.

8. Kelvin, *Baltimore Lectures*, 260–278. His treatment of the crux of the night sky puzzle is also reprinted in Harrison, *Darkness at Night*, 227–228.

Chapter 10 The Perfect Blue

1. As Max von Lane notes, "unlike the ear, which harmonically analyzes the vibrations that are stimulating it, [the eye] perceives white light as a unit"; *History of Physics*, tr. Ralph Oesper (New York: Acadenic, 1950), 34.

2. Kelvin, *Baltimore Lectures*, 302.

3. See *Optical Papers of Isaac Newton*, ed. Alan E. Shapiro (Cambridge: Cambridge University Press, 1984), 1:537–549, at 545, 50 note 10, K. McLaren, "Newton's Indigo," *Color Research and Application* 10(4), 225–229 (1958), and Sepper, *Newton's Optical Writings*, 95–99, 205–207.

4. For instance, consider the shining violet eyes of Miss Turner in "The Boscombe Valley Mystery," *Annotated Sherlock Holmes* 2:140.

5. For the colors in the rainbow, see Boyer, *Rainbow*, 47–57, 101, 239–244; see also Newton, *Opticks*, 37, 48–52.

6. A. Crova, "Sur l'analyse de la lumière diffusée par le ciel," *Annales de Chimie Physique* 20, 480–504 (1890), at 487.

7. For the history of skylight spectra, see Kelvin, *Baltimore Lectures*, 306–323, and S. T. Henderson, *Daylight and Its Spectrum* (New York: American Elsevier, 1970), 58–74, 203–208. Among experimental studies, particularly significant are W. de W. Abney, "On the Colours of Sky Light, Sun Light, Cloud Light, and Candle Light," *Proceedings of the Royal Society of London* 54, 2–4 (1893), which measures the dominant wavelength of skylight; A. Bock, "Der blaue Dampfstrahl," *Annalen der Physik und Chemie* 68, 674–687 (1899), which makes detailed comparison with light scattering from steam in the laboratory; Quirino Majorana, "On the Relative Luminous Intensities of Sun and Sky," *Philosophical Magazine*, 555–562 (May 1901), which follows Kelvin's suggestion to compare sunlight and skylight at high altitude; and Giuseppe Zettwuch, "Researches on the Blue Colour of the Sky," *Philosophical Magazine* 4, 199–202 (1902), which makes comparisons of skylight, lamplight, and light from steam scattering on wavelengths into the violet.

8. Edward L. Nichols, "Theories of the Color of the Sky," *Proceedings of the American Physical Society* 6, 497–511 (1908), from which his figure 11 has been taken as my figure 10.1, and "A Study of Overcast Skies," *Physical Review* 28, 122–131 (1909).

9. Henderson, *Daylight*, 61–64.

10. *The Science of Color*, ed. Steven K. Shevell, second ed. (Boston: Elsevier, 2003), 44–45, 91 gives the CIE data used in figure 10.4. See also Glenn S. Smith, "Human Color Vision and the Unsaturated Blue Color of the Daytime Sky," *American Journal of Physics* 73, 590–597 (2005).

11. See Edwin H. Land, "Experiments in Color Vision," *Scientific American* 200(5), 84–99 (1959).

12. Cambridge University MS. Add. 3975, 15.

13. See "The Case of the Colorblind Painter," in Oliver Sacks, *An Anthropologist on Mars: Seven Paradoxical Tales* (New York: Knopf, 1995), 3–41,

especially 18–31, on the roles of the V1 and V4 centers in color perception.

14. For example, see Bernard H. Soffer and David K. Lynch, "Some Paradoxes, Errors, and Resolutions Concerning the Spectral Optimization of Human Vision," *American Journal of Physics* 67, 946–953 (1999); James M. Overduin, "Eyesight and the Solar Wien Peak," *American Journal of Physics* 71, 216–219 (2003); and Mark A. Heald, "Where Is the 'Wien Peak'?" *American Journal of Physics* 71, 1322–1323 (2003).

15. See N. A. Voishvillo and Yu. A. Anokhin, "Color of a Plane Layer of a Medium Scattering According to Rayleigh Law," *Izvestiya, Atmospheric, and Oceanic Physics* 20, 760 (1984), and Craig F. Bohren, "Multiple Scattering of Light and Some of Its Observable Consequences," *American Journal of Physics* 55, 524–533 (1987).

16. Aristotle, *Meteorology*, 372a29–372b8.

17. Craig F. Bohren, "Multiple Scattering at the Breakfast Table," *Weatherwise* 36(3), 1443 (1983), and again in his "Multiple Scattering of Light" and his *Clouds in a Glass of Beer*, 104–119.

18. The number of scatterings depends on the mean free path for light in air, $l = 1/\sqrt{2}n\sigma$, where n is the number density of air molecules (about 3×10^{25} molecules/m^3 at sea level) and σ is the scattering cross section, the "effective area" of the scatterer. A medium is "optically thick" if it contains many mean free path lengths of light. The Rayleigh scattering cross section is $\sigma = (8\pi/3)(e^2/mc^2)^2\lambda_0^4/(\lambda^2 - \lambda_0^2)^2$, where e and m are the charge and mass of the electron and c the speed of light; $e^2/mc^2 = 2.8 \times 10^{-15}$ m is the "classical electron radius," the only combination of e, m, and c that has the dimensions of length. Rayleigh scattering needs *bound* electrons; free electron scattering is independent of wavelength. λ_0 is a resonant wavelength for atmospheric molecules, typically in the ultraviolet, say 200–240 nm. These values yield mean free paths of 1–3 km for violet light ($\lambda = 400$ nm) at sea level. The mean free path is inversely proportional to the atmospheric pressure and so grows with altitude. If the whole atmosphere were constrained to have uniform density, it would have a *scale height* of about 8 km; this gives a measure of the "thickness of the atmosphere" that is comparable to the

mean free path, showing that most light coming from the zenith is scattered only once: Our atmosphere is "optically thin." In contrast, light coming obliquely from the horizon passes through a much longer path and thus is likely to experience multiple scattering.

19. See Bohren and Fraser, "Colors of the Sky," 271–272, and the discussion of the volcanic purple light in Raymond L. Lee, Jr. and Javier Hernández-Andrés, "Measuring and Modeling Twilight's Purple Light," *Applied Optics* 42, 445–457 (2003).

20. E. O. Hulburt, "Explanation of the Brightness and Color of the Sky, Particularly the Twilight Sky," *Journal of the Optical Society of America* 43, 113–118 (1953) (in Bohren, *Scattering*, 75–80). See also the papers collected in Bohren and Huffman, *Absorption and Scattering of Light by Small Particles*.

21. Charles N. Adams, Gilbert N. Plass, and George W. Kattawar, "The Influence of Ozone and Aerosols on the Brightness and Color of the Twilight Sky," *Journal of Atmospheric Sciences* 31, 1662–1674 (1974). See also G. V. Rozenberg, *Twilight: A Study in Atmospheric Optics*, tr. R. G. Rodman (New York: Plenum 1963).

22. See Jerry L. Holechek, Richard A. Cole, James T. Fisher, and Raul Valdez, *Natural Resources: Ecology, Economics, and Policy* (Upper Saddle River, N.J.: Prentice Hall, 2000), 175. There is a very informative Web site presentation at http://www.atm.ch.cam.ac.uk/tour/part1.html.

23. See Meinel, *Sunsets, Twilights, and Evening Skies*, 127–131.

24. Henry James: "The Next Time," in *The Novels and Tales of Henry James* (New York: Augustus James, 1970), 216.

25. Isak Dinesen, "The Young Man with the Carnation," from *Winter's Tales* (New York: Random House, 1993), 18–20.

Appendix A Experiments

The best collections of home experiments in atmospheric physics are found in Bohren, *Clouds in a Glass of Beer* and *What Light through Yonder Window Breaks?*

Appendix B Letters on Sky Blue between George Gabriel Stokes, John Tyndall, and William Thomson

1. [These hitherto unpublished letters are from the collections of the Royal Institution of Great Britain, whom I thank for permission to print them here and whose archivist, Lenore Symons, has been extremely gracious and helpful. I especially thank David B. Wilson for his generous help deciphering difficult passages in the holographs and for supplying me with copies of letters between Tyndall and Stokes. I have quoted several of the letters in their entirety; where I have summarized or commented, my remarks are enclosed in square brackets.]

2. [See Stokes, "On the Change of Refrangibility of Light" and Hey, "From Leonardo to the Graser," I, 16–21.]

3. [Note Stokes' hesitation to dismiss the possibility of "molecular reflection," compared with Tyndall's outright rejection. Stokes here he alludes to the crucial physical criterion, essentially the optical thickness of the atmosphere; see chapter 10, pp. 241–242, note 18.]

4. [Roscoe, "On the Opalescence of the Atmosphere." Using sensitive photographic paper, Roscoe measured the chemical action of skylight at different times of year and in different places, including Kew Observatory, Owen's College, Manchester, and Pará, Brazil. I have mentioned his experiments on light scattering from dilute sulfur suspensions in chapter 6, pp. 96–97. Note that Stokes here rather pointedly draws Tyndall's attention to this work as importantly anticipating some of his experiments. Tyndall later mentioned Roscoe's paper in his essay "Climbing in Search of the Sky," 15, though without discussing the relation between their respective experiments.]

5. [Brücke cannot justify to Stokes that light of different wavelengths could be *reflected* differently, since specular reflection acts equally on all colors; Tyndall will resolve the impasse by arguing that *scattering* can act preferentially on shorter wavelengths.]

6. [For Thomson's work on determining molecular dimensions, see chapter 7, pp. 230–231, note 10. For his and Stokes' changing opinions

about "ultimate molecules," see Wilson, *Kelvin and Stokes*, 107–113, 16–169, which also cites Stokes's 1878 correspondence with William Crookes, who argued that "the blue colour of the sky is the colour of phosphorescence" (198–199).]

7. [In *Fragments of Science*, 119–122, Tyndall details these microscopic examinations and gives as an upper limit of the precipitated particle size 1/100,000 inch (100 nm).]

8. [To explain the source of the sun's energy, Kelvin had speculated that it came from the release of gravitational potential energy supplied by infalling meteors. See above, p. 156, and Burchfield, *Lord Kelvin and the Age of the Earth*, 134–162.]

9. [A bacterium with a curled, rodlike shape; some writers used the term "vibrio" interchangeably with "bacterium." See Strick, *Sparks of Life*, 198.]

Acknowledgments

To Larry Cohen, Sara Meirowitz, Judy Feldmann, and their associates at the MIT Press, whose outstanding support and collaboration helped bring this book to life.

To St. John's College, where I learned how deep is the question: Do atoms really exist?

To Craig F. Bohren, Michael Vollmer, Andrew Young, and Curtis Wilson, whose criticisms saved me from many mistakes. The errors that remain are my own.

Finally, to Ssu, Andrei, and Alexei, who mean more to me than the sky.

Index